AS/A-LEVEL

STUDENT GUIDE

WJEC/Eduqas

Geography

Changing places

David Burtenshaw and Kevin Davis

This Guide has been written specifically to support students preparing for the WJEC/Eduqas AS/A-level Geography examinations. The content has been neither approved nor endorsed by WJEC/Eduqas and remains the sole responsibility of the authors.

Although every effort has been made to ensure that website addresses are correct at time of going to press, Hodder Education cannot be held responsible for the content of any website mentioned in this book. It is sometimes possible to find a relocated web page by typing in the address of the home page for a website in the URL window of your browser.

Hachette UK's policy is to use papers that are natural, renewable and recyclable products and made from wood grown in well-managed forests and other controlled sources. The logging and manufacturing processes are expected to conform to the environmental regulations of the country of origin.

Orders: please contact Hachette UK Distribution, Hely Hutchinson Centre, Milton Road, Didcot, Oxfordshire, OX11 7HH. Telephone: (44) 01235 827827. Email: education@hachette.co.uk. Lines are open from 9 a.m. to 5 p.m., Monday to Friday. You can also order through our website: www.hoddereducation.co.uk.

ISBN 978-1-5104-7216-7

First printed 2020

First published in 2020 by
Hodder Education,
An Hachette UK Company
Carmelite House
50 Victoria Embankment
London EC4Y 0DZ

www.hoddereducation.co.uk

Impression number 10 9 8 7 6 5 4 3 2

Year 2024 2023 2022 2021

Cover photo: Mihai Andritoiu/Adobe Stock

Typeset by Integra Software Services Pvt. Ltd, Pondicherry, India

Printed and bound by CPI Group (UK) Ltd, Croydon, CR0 4YY

A catalogue record for this title is available from the British Library.

Contents

Content Guidance

Questions & Answers

Getting the most from this book

Exam tips

Advice on key points in the text to help you learn and recall content, avoid pitfalls, and polish your exam technique in order to boost your grade.

Knowledge check

Rapid-fire questions throughout the Content Guidance section to check your understanding.

Knowledge check answers

1 Turn to the back of the book for the Knowledge check answers.

Summaries

- Each core topic is rounded off by a bullet-list summary for quick-check reference of what you need to know.

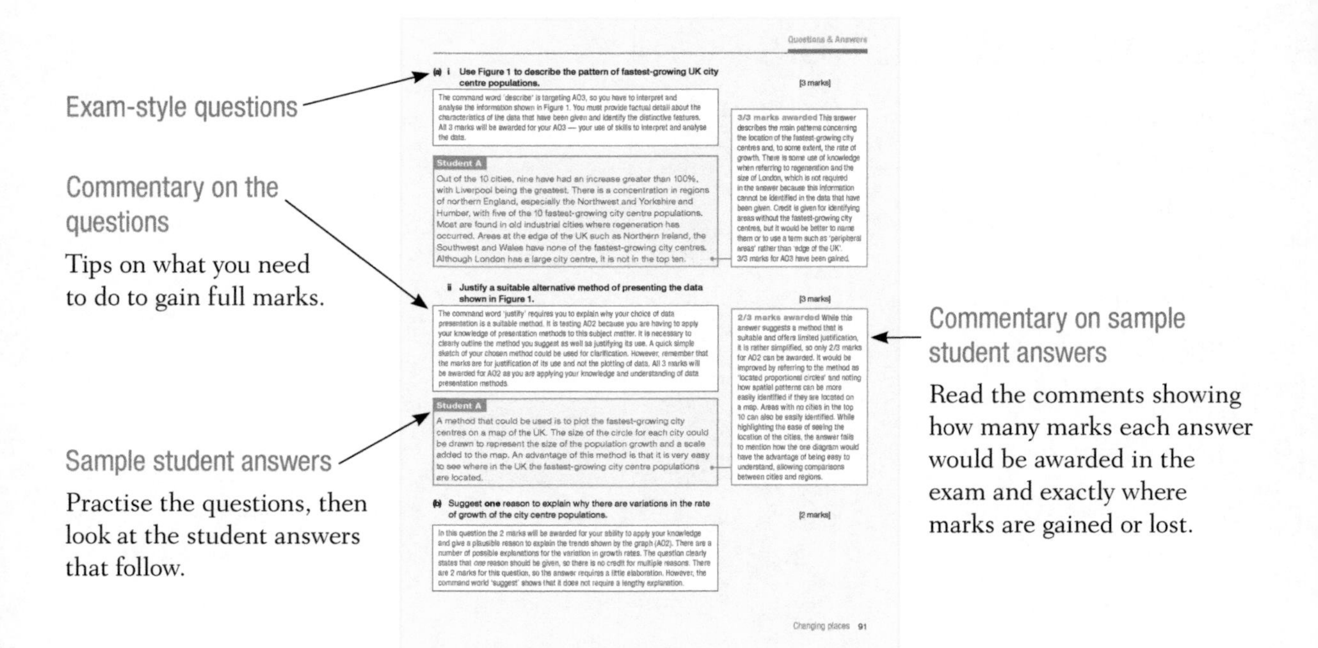

Questions & Answers

(a) i Use Figure 1 to describe the pattern of fastest-growing UK city centre populations. [3 marks]

The command word 'describe' is targeting AO3, so you have to interpret and analyse the information shown in Figure 1. You must provide factual detail about the characteristics of the data that have been given and identify the distinctive features. All 3 marks will be awarded for your AO3 — your use of skills to interpret and analyse the data.

Student A

Out of the 10 cities, nine have had an increase greater than 100%, with Liverpool being the greatest. There is a concentration in regions of northern England, especially the Northwest and Yorkshire and Humber, with five of the 10 fastest-growing city centre populations. Most are found in old industrial cities where regeneration has occurred. Areas at the edge of the UK such as Northern Ireland, the Southwest and Wales have none of the fastest-growing city centres. Although London has a large city centre, it is not in the top ten.

3/3 marks awarded This answer describes the main patterns concerning the location of the fastest-growing city centres and, to some extent, the rate of growth. There is some use of knowledge when referring to regeneration and the size of London, which is not required in the answer because this information cannot be identified in the data that have been given. Credit is given for identifying areas without the fastest-growing city centres, but it would be better to name them or to use a term such as 'peripheral areas' rather than 'edge of the UK'. 3/3 marks for AO3 have been gained.

ii Justify a suitable alternative method of presenting the data shown in Figure 1. [3 marks]

The command word 'justify' requires you to explain why your choice of data presentation is a suitable method. It is testing AO2 because you are having to apply your knowledge of presentation methods to this subject matter. It is necessary to clearly outline the method you suggest as well as justifying its use. A quick simple sketch of your chosen method could be used for clarification. However, remember that the marks are for justification of its use and not the plotting of data. All 3 marks will be awarded for AO2 as you are applying your knowledge and understanding of data presentation methods.

Student A

A method that could be used is to plot the fastest-growing city centres on a map of the UK. The size of the circle for each city could be drawn to represent the size of the population growth and a scale added to the map. An advantage of this method is that it is very easy to see where in the UK the fastest-growing city centre populations are located.

2/3 marks awarded While this answer suggests a method that is suitable and offers limited justification, it is rather simplified, so only 2/3 marks for AO2 can be awarded. It would be improved by referring to the method as 'located proportional circles' and noting how spatial patterns can be more easily identified if they are located on a map. Areas with no cities in the top 10 can also be easily identified. While highlighting the ease of seeing the location of the cities, the answer fails to mention how the one diagram would have the advantage of being easy to understand, allowing comparisons between cities and regions.

(b) Suggest one reason to explain why there are variations in the rate of growth of the city centre populations. [2 marks]

In this question the 2 marks will be awarded for your ability to apply your knowledge and give a plausible reason to explain the trends shown by the graph (AO2). There are a number of possible explanations for the variation in growth rates. The question clearly states that *one* reason should be given, so there is no credit for multiple reasons. There are 2 marks for this question, so the answer requires a little elaboration. However, the command world 'suggest' shows that it does not require a lengthy explanation.

Changing places 91

About this book

This Guide has been designed to help you succeed in the Eduqas A-level and WJEC AS and A-level Geography **Changing places** topics. The guide has two sections.

The **Content Guidance** summarises the key information that you need to know to be able to answer the examination questions with accuracy and depth. In particular, the meanings of key terms are made clear. You will also benefit by testing your knowledge with knowledge check questions, and noting the exam tips, which provide further help in determining how to learn key aspects of the course.

The **Questions & Answers** section includes sample questions similar in style to those you might expect in the exam. There are sample student responses to these questions as well as detailed commentary giving further guidance in relation to what exam markers are looking for in order to award top marks. The best way to use this book is to read through the relevant topic area first before practising the questions. Only refer to the answers and comments after you have attempted the questions.

The topics covered in this guide make up:

- **Section B of Eduqas A-level Geography Component 1**
- **Section A of WJEC AS Geography Unit 2**

The formats of the different examination papers are summarised in the table below.

Specification and paper	Total time for Changing places	Total marks for Changing places	Structured questions	Extended response/essay
Eduqas A-level Component 1, Section B	**50 min** in a paper lasting 1 h 45 min	41/82	**Two** compulsory, structured, data-response questions marked out of 13	**One** compulsory question marked out of 15
WJEC AS Unit 2, Section A	**45 min** in a paper lasting 1 h 30 min	32/64	**Two** compulsory structured, data-response questions marked out of 16	None

Content Guidance

Changing places: relationships and connections

The characteristics of places

On your journey to college or school, what types of urban or rural area do you pass through? One student may depart on a bus from a small development of 1980s houses, a small community, which the local residents' committee call a **village**, despite it being in an **exurb** of 17,000 people swallowed up by suburbia. Her journey takes her through an area of semi-detached homes built between 1920 and 1939, the residents of which voted solidly Conservative in the 2019 general election, and past a retail park. The bus enters a declining traditional **suburban** high street with small retailers, estate agents, charity shops, betting shops and a bingo hall in the former cinema. This 'high street' locality also houses some local government offices. Her route continues past 'motor row' — car sales showrooms on former military land, some of which is the location for new high-tech industries. Next, through nineteenth-century terraced housing, with its corner shops, pubs turned into supermarket chain stores and a cinema converted into a mosque. Some of her friends from the South Asian communities board the bus. They are worried as some people in this **neighbourhood** voted for the UK Independence Party (UKIP).

The bus takes the inner ring around the city centre, with its landscape of car parks surrounding the pedestrianised shopping area where some shops and a department store have closed. She finally reaches her school, on the fringe of the old medieval town centre, abandoned as the central area over 100 years ago. This is the desirable part of the city (**neighbourhood**) to live in (**identity**) and it is close to the local university.

What characterises the places she identified? She identifies architecture (**landmarks**), the areas' cultural, social and demographic mix, and their economic profile. She is also recalling her own mental map based on the **pathway** that her journey takes. She has identified **edges** between **areas** where there is a defined change, such as around the city centre. She notes that places are dynamic because they adapt and change over time.

The characteristics of a place can therefore be influenced by many factors:

- Demographic — the size and structure of the population.
- Socioeconomic — this includes the incomes, types of employment and education opportunities. It also includes the health of the population and crime rates in the area.
- Cultural — the religions, customs, languages and social behaviour that can make a place distinct.

- Built environment — the land use, including the level of urbanisation, building types and density.
- Physical geography — the relief and physical features, such as rivers and coasts, which can add to a place's character.

People use different terms for the places we live in. Some of these key definitions are as follows:

- **Built-up area:** defined by the 2011 census as areas of built-up land that are joined together, and where the gaps between the developed areas are less than 200 m. For example, Cardiff's built-up area includes Penarth, Pontypridd and Caerphilly.
- **City:** a large settlement depending primarily on service and knowledge industries, together with manufacturing. It is an aggregation of places. In 2015 UK cities made up 9% of the land area of the UK and yet housed 54% of the population. 59% of jobs and 72% of the highly skilled workers live in them. 78% of new migrants live in cities. Officially, cities in the UK have been granted city status by royal charter. There are currently 69, of which 51 are in England and six in Wales. Some still cling to the idea that a city has to have a cathedral. However, these are historical definitions rather than definitions based on current functions, and hardly apply to many other countries.
- **City region:** an area served by and functionally bound to a city, and normally including journeys to work and to places of study.
- **Community:** a set of interacting but diverse groups of people found in a particular locality. It may be tied together by common heritage, but many communities can be very diverse.
- **Conurbation:** an urban area that has fused together over time, such as Greater Manchester or the Ruhr region of Germany. It may grow from one centre (e.g. London) or from several (e.g. the West Midlands).
- **Dispersed city:** a US term used to define a city that has sprawled over a wide area, such as the spread of the San Francisco urban area around San Francisco Bay and up to 60 miles beyond into California.
- **Dormitory village and commuter village:** used to describe a settlement in which the population is socially urban and works in nearby urban areas. Estate agents frequently misappropriate the term to describe new developments and boost the attractiveness of neighbourhoods within cities.
- **Exurbia:** those areas beyond the urban area that house people who live in mainly rural surroundings but work in urban areas.
- **Global hub:** a large city that is at the heart of the global economic and financial system, for example London, Tokyo, Shanghai, New York.
- **Hamlet:** a small cluster of dwellings/farms that lacks services.
- **Isolated dwelling:** a single rural dwelling (or pair of dwellings), often in a sparsely populated area.
- **Locality:** descriptive term for where people live out their daily working and domestic lives. It can vary in size — geographers use the term loosely for anything from small scale to a large urban area.
- **Megacity:** a city with a very large population, for example Shanghai, Tokyo, Mexico City.
- **Megalopolis:** a growing together of large urban and suburban areas, for example Boswash, the area between Boston and Washington DC, USA.

Exam tip

It is essential that you support your studies with examples from contrasting places. Make copious use of your 'home' place, which may be a locality, neighbourhood or small community, as well as the region in which it is located. Remember that you can use examples from field studies.

Exam tip

You need to know and use the various place/settlement terms and definitions that are used by geographers.

- **Metropolitan area:** a term frequently used instead of conurbation.
- **Minor built-up rural area:** a rural area with a main settlement of under 10,000 people.
- **Neighbourhood:** a distinct and recognisable residential area that may be the location of home and its immediate environment. It can be someone else's home base and area. It is an area of similar housing, persons and lifestyles. The term was used most obviously in the planning of new towns in the 1960s, having originated in the Garden City movement.
- **Primary urban area (PUA):** a built-up area of a city that invariably extends beyond its administrative area. The term is used in publications of www.centreforcities.org.
- **Rural settlement:** a village, hamlet and/or isolated farms in the countryside formerly associated with primary employment. Most of the population of rural settlements do not work in the countryside. Rural UK is psychologically urban because the hamlets and villages contain people who may have retired from city jobs, or who work in nearby urban areas. Peter Hall stated in 2014 that there are few truly rural settlements within 150 miles of London.
- **Rural–urban fringe:** a dated term that refers to the immediate surroundings of an urban area, which contain elements of an urban area, such as golf courses alongside open countryside.
- **Suburb:** an area of mainly residential units that has been developed around the core of a town or city. It has increasingly included other uses, such as industry, retailing, offices, recreational buildings and public open spaces.
- **Town:** a small urban area with a range of services that may include some independent retailers, schools (sometimes secondary), post offices, banks and estate agents.
- **Village:** a small rural settlement with some functions, for example a post office, shops, public houses and a church. Population sizes vary (200–7,000 in the UK, but far larger in Italy, for example).
- **World city:** a broader term than global hub that refers to a city that is prominent in the operation of the world economy, for example Hong Kong, Singapore, Frankfurt, Paris.

All of these terms are not entirely discrete because communities and localities are embedded in the suburbs of cities or in villages. They nest within larger places and give those places a **meaning** because of the combination of localities or communities.

What makes a place distinct?

Arundel in West Sussex is a small former market town or large village. In 1960 the population was around 2,600. Living in the Arundel of 1960 was different (yet also similar in many ways) from living in the Arundel of 2019 (**time**). The shops catered for local people and were locally owned. Greengrocers, butchers, bakers, a dairy, 'ladies and gents outfitters', two banks, a jewellers, a cinema and some small factories catered for the needs of the population. The Arundel Castle Estate, its farms and horse racing stables, provided employment for many, although some did commute by bus and train to London and nearby towns. There were many public houses and a former coaching inn. A council estate had just been completed on the fringe of the town (**identity**). The population had declined in the previous decade and 21% of the inhabitants were retired.

Arundel (population 3,295 in 2017) now functions as a tourist centre within the South Downs National Park. A more commercialised castle, a cathedral, jailhouse, museums and the Wildfowl and Wetlands Trust nature reserve are among its attractions. Its **identity** and **representation** changed through modernisation. The centre has many antique shops and small galleries, an antique bookshop and specialist food shops, together with a range of bars, the same coaching inn, and restaurants and cafés catering for all tastes, especially those of visitors.

However, its two banks have closed. The **threshold** population required to sustain their services has grown to the point that local retailing and banking are no longer viable. A farmers' market has returned to the town, which is a Fair Trade town. Much of the population now commutes to work in larger towns by car, and some by rail to London. The retired population of the town had increased to 31% by 2017. The Office for National Statistics classifies Arundel as a 'coastal and countryside senior community' because its demography resembles nearby coastal places, such as Littlehampton and Bognor Regis, and countryside places such as Petworth and Midhurst. Although it has changed as the economy has altered to a service economy due to the growth of tourism, the identity of the main street and most of the streets in the old town has not changed. However, while the buildings are identical, their functions have changed. Therefore, the differences in 2019 are not architectural but social and economic. Nevertheless, the architecture enables the town to continue to function as a tourist centre — it is **resilient**.

Exam tip

Make sure that you can categorise your home place and are able to identify the different factors that give your home area a distinct character.

Clovelly in Cornwall has changed its economic base, and yet its visual image remains the same. Once primarily a small mackerel fishing harbour at the mouth of a steep valley, with houses clinging to the valley sides and cliff edge, the village has become a tourist hot spot that is still **represented** by the village's personality as a harbour. Mackerel fishing, once strictly controlled by EU fishing policy, has declined to just one boat, but tourism has boomed. External forces have changed Clovelly's economic base, but the image of the place remains **resilient**.

Continuity and change in contrasting places

The Lake District

Place can be a larger area that can be seen in different ways by different people.

Exam tip

Remember that 'place' does not only refer to an urban area. It could refer to remote rural areas such as parts of the Lake District.

The characteristics of a place can be shaped by many factors that may change over time. Despite the relief of the land and wet climate creating difficulties for agriculture, there has been a tradition of farming in the Lake District for thousands of years. Generations of the same family have farmed areas for hundreds of years. In 1851 around 40% of the population worked in agriculture.

However, in early Victorian times the Lake District was already becoming a popular tourist destination, helped by the writing of artist and social revolutionary John Ruskin, who described the area as 'The loveliest rock scenery, chased with silver waterfall, that I have ever set foot or heart upon.' William Wordsworth's *Guide to the Lakes*, first published in 1810, also encouraged visitors to the area, with still more after the arrival of the railways. It was during the nineteenth century that the Lake District economy shifted from being dependent on agriculture to being more centred on tourism. In 2017, 19.17 million people visited the Lake District, spending £1.4 billion and helping to create 18,500 jobs. Today, of the 41,000 living in the park, 2,500 work in agriculture and 15,000 in tourism.

The population continues to change. The 64+ age group is increasing, especially among the more affluent looking for a quality of life. Lifestyle, education, employment and housing opportunities have all resulted in a decline in the 16–30 age group. By 2029 over 50% of the population will be over 50. Around 15% of all dwellings are now holiday or second homes.

In July 2017, the Lake District became a World Heritage Site. The nomination document noted that 'The harmonious beauty of the Lake District is rooted in the interaction between an agropastoral land use system and the spectacular natural landscape of mountains, valleys and lakes'. However, the writer George Monbiot writing in the *Guardian* at the time felt the designation turned the Lake District into a 'Beatrix Potter-themed sheep museum', with '...entire high fells reduced by sheep to a treeless waste of cropped turf, whose monotony is relieved only by erosion gullies, exposed soil and bare rock'.

Foster City, California

A **contrasting place** to the Lake District and probably your home location is Foster City, California (Figure 1). Foster City is an affluent, planned settlement of 34,412 people (2019) built on reclaimed marshland on the western shore of San Francisco Bay. Jack Foster bought Brewer Island and the surrounding wetlands in 1958, drained it and raised its level by dredging 10.7 m^3 of sands and muds from the Bay.

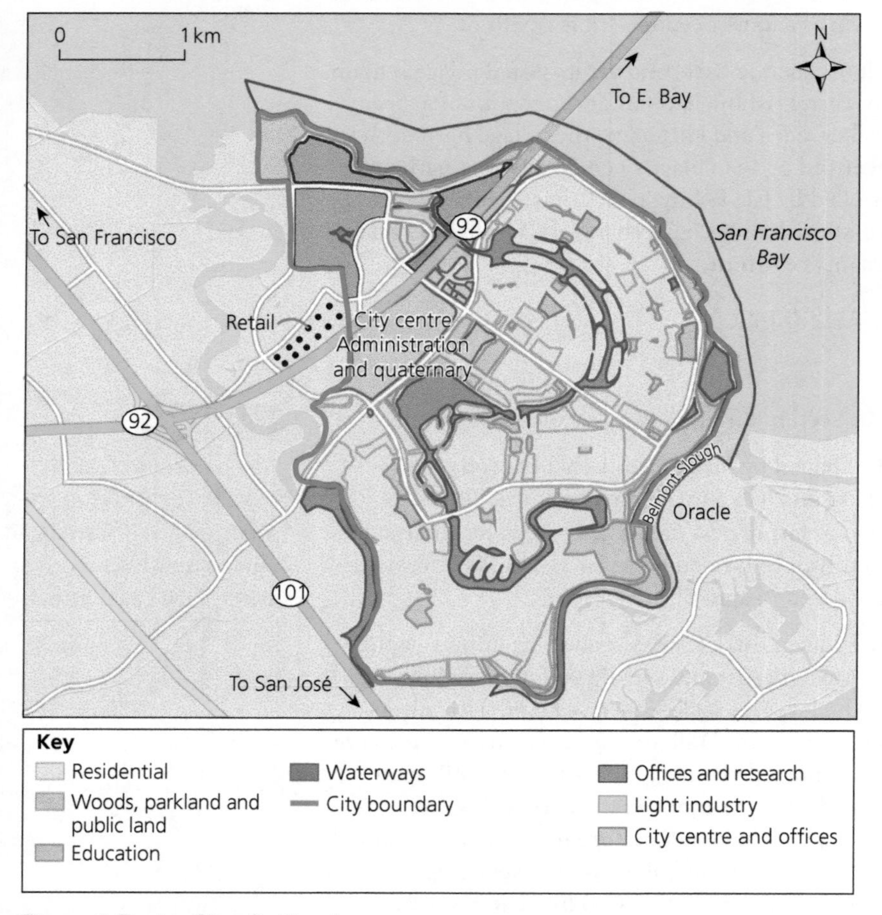

Figure 1 Foster City, California

Knowledge check 1

What are the socioeconomic and demographic factors that have resulted in changes in the characteristics of the Lake District?

Exam tip

You need to have knowledge of at least one contrasting place. You could contrast the example above with your home place.

Today the place is distinct because it is an urban area defined and given identity by the Bay and Belmont Slough, a tidal watercourse. The city covers 50 km^2 but almost 40 km^2 of the area comprises seawater lagoons and the Slough, created when the city was first developed. The local people navigate these with small (electric, sail or human-powered) vessels.

The 'city' is primarily a medium-to-high-earner residential community (Figure 1) although, with the passage of time, the proportion of the retired population is growing. There is no city centre in a European sense — it only houses the city administration and regional offices for Visa and IBM. There are several small retail areas, but the main shopping is in malls either at the edge of, or beyond, the city limits. Employment is often elsewhere, leading to a daily outflow of commuters to Silicon Valley headquarters (page 52). Quaternary work and the knowledge economy dominate, as firms move into the area and cluster near the main road access routes from the dispersed Bay City. By 2017, 16,000 people worked here, many commuting in from cities in the Bay area. Almost 26% of the population work in professional, scientific and technical services, compared with an average of 6.7% for the USA.

The population composition enhances the uniqueness (**identity**) of Foster City. In 2017, 48% were of Asian origin, the majority of whom were born in China (including Hong Kong), India, Japan and Taiwan (**interdependence**). They have the highest average earnings. The white US and European population is 41% of the total (20% were born outside of the USA and only 38% born in California). A further 6.5% are Hispanic, a far lower proportion than in the rest of California. The African-American population is 2.1% of the total; on average, they are the lowest earners. Foster City is a city of immigrants with a multicultural identity. The cost of living here is 2.5 times that of the US average. The average household income is US$130,000 — over double the US average. House prices average US$1.6m, which is three times the average for California.

The role of iconic objects

The identity of a place can sometimes be defined by an **object**, such as an iconic building or a work of art. The 'Google bicycle' refers to a fleet of over 1,000 bicycles produced in the colours of Google's logo which are used by 13,000 employees at the Mountain View California HQ to cycle between its buildings. The buildings' functions are not posted outside and therefore the parked bicycles (**representation**) give a spatial extent to the 'Google campus'. A Google journey-to-work region can also be defined by the extent of the routes used by the frequent free Google buses, which bring employees to the site.

Knowledge check 2

Suggest examples of places in the UK that are frequently identified or associated with an iconic object or building.

Factors that shape the changing identity and characteristics of places

Table 1 lists the factors that affect the changing identity of places ranging from villages and neighbourhoods to regions. These factors may be changes in the flows and connections between people, resources, money and ideas.

Table 1 Factors that affect the changing identity of places

Factors	Village	Neighbourhood	Suburbs	City centre	Region
Cultural	New economy	Community-based	Segregation	Leisure quarter	—
Economic	Decline of primary economy and rise of service employment	Adaptation to new or existing demographic	Industrial estates/ office parks	Tertiary and quaternary growth; R&D	Deindustrialisation; global changes, e.g. the web, multinational companies
Investment	Global investors; new rural economy	Renovation	Renovation	New leisure services and retail decline	Inward investment/ Foreign Direct Investment (FDI), infrastructure investment
Resources	New land uses	Community-based; specialist retail	—	Knowledge industries; universities	Accessibility nationally and internationally
Demographic	Ageing population	Homogeneity	Ageing, diverse communities	Reurbanisation, gentrification	Unemployment, deprivation
Migration	Young leave and mature move in	Nest builders	Life cycle moves; social segregation	Student flats	In- and out-migration
Planning	Expanded settlement; counterurbanisation	Garden City; neighbourhood units	Green belts	Redevelopment	Development grants
Political	NIMBYism	Neighbourhood Watch	—	Rebranding/ boosterism	Government Regional Policies
Global forces	Global agricultural production	Satellite television	Retail/fast food, e.g. ASDA and McDonald's	Global banks, financial companies, leisure chains; private involvement	FDI

Many multinational corporations (MNCs) are involved in a wide variety of activities and retailing in cities. This investment can lead to a decrease in the uniqueness of some town centres as they become 'clone towns', where over 60% of the shops are recognisable chain stores. This impact can even be global, with identical chain stores, fast-food restaurants and products being recognisable in many countries (Table 2).

Knowledge check 3

How does Table 2 show that the interdependence of towns is influencing their character?

Table 2 MNC ownership of some well-known shops

Name	Type of activity	No. of outlets in UK	No. of outlets worldwide	No. of countries where it operates	Where owned
McDonald's	Fast food	1,249	37,855	120	Oak Brook, Illinois, USA
KFC	Fast food	842	23,000	119	Louisville, Kentucky, USA
Domino's Pizza	Fast food	1,100	16,000	85	Ann Arbor, Michigan, USA
Pizza Express	Restaurant	490	640	13	Hony Capital, Hong Kong, China
Starbucks	Coffee shop	335	29,324	76	Seattle, USA
Costa Coffee	Coffee shop	2,389	3,755	31	Dunstable, UK (Part of Coca-Cola Company)
Aldi	Discount supermarket	775	11,234	20	Essen, Germany
Lidl	Discount supermarket	710	10,500	30	Neckarsulm, Germany
Arcadia Group	Clothing (Topshop, Topman, Burton, Dorothy Perkins, Evans, Miss Selfridge, Wallis, Outfit)	2,500	3,760	37	London, UK
Waterstones	Bookshop	275	283	4	Elliott Management, New York, USA

How does continuity and change affect lives?

Continuity or change in any of the factors that shape the characteristic of a place can impact on the lives of the people who live there. Bermondsey, south of the River Thames and situated between London Bridge and Tower Bridge, has experienced many changes that affect lives. The poor inhabited the area even in Shakespeare's time, when the area straddled the road south from the only Thames crossing. It had markets, e.g. Borough Market, wharves along the Thames, which were London's sole docks, and 'noxious industries', as well as tanning, taverns, brothels, coaching inns and slums, which inspired Charles Dickens's *Oliver Twist*.

By the nineteenth century rail routes crossed the area (which, in the twentieth century, now enclose areas of municipal housing that replaced some of the slums). Industry grew, including a gas works, printing and food processing of goods imported through the port.

In the Second World War, much of the area was destroyed during the Blitz. These areas then became a focus for council housing as a result of comprehensive renewal (**causality**). The docks grew derelict because of new shipping technologies further downstream.

In the 1980s, the London Docklands Development Corporation (LDDC) began to redevelop the area. Commercial developments were housed in converted buildings and private investment attracted a middle-class population into converted warehouses. On the western fringes, the Globe Theatre (1996) and Tate Modern (2000) became part of the South Bank cultural complex. Jobs in the traditional industries declined, replaced by public services and creative industries. By 2011, the population was more cosmopolitan and socially polarised in one of the most deprived areas of the capital (**adaptation** and **time**).

Today, the population is younger and there are many more residents with higher education qualifications than the national average. In 2018, Bermondsey was voted the best place to live in London by the *Sunday Times*.

Serena, an unemployed mother living on a Bermondsey council estate (**community**), states: 'Everybody on the estate knows everybody else. The estate is hospitable because it is small (**attachment**). Other estates don't have community because they are tall blocks. People tend not to give up their flats on this estate.'

In contrast, Henrietta, a management software consultant, lives in a warehouse conversion at Butler's Wharf: 'I love it here — going for a coffee, to the bakery and a drink at All Bar One, and to the good restaurants (**identity**). The flats are really expensive, but you would not talk to the people in the lift.'

These two experiences illustrate how change in the area has affected people's lives in different ways.[1]

Events and decisions at a global level can affect people at a local level

In the past, decisions made at a national scale affected people, whereas today, global-level decisions affect local people. Worldwide **interdependence** occurs due to the nature of the modern global economy, trade patterns and communications.

[1]The quotes are adapted from Hall, P. (2007) *London Voices, London Lives*, The Policy Press. The text contains many interviews about changing places across London.

In 2015, Volkswagen (VW) was found to have breached regulations concerning emissions from cars. This resulted in fewer people purchasing VW vehicles. The main VW car plant in Wolfsburg, Germany employs 70,000 workers, all of whom received a profit bonus of €5,900 (£4,900) in 2014. The bonus did not occur in 2015 because the firm was forced to cover the costs of selling fewer cars. In 2016 the company restructured and cut the number of employees, many at its Wolfsburg factory. Business in the city that depended on VW employees' spending also had to readjust; for example, osteopaths and chiropractors forecast a 25% drop in patient numbers and income. Redundancy hit shops, supply firms and other services. Workers took fewer holidays, which had an impact on the local tourism economy. The communities in and around Wolfsburg were also affected.

Global technological changes have allowed people to change the way they shop, which has had consequences for people at the local level. Each year, the John Lewis Partnership pays its staff a bonus. In 2019 the bonus was the lowest since 1953 as a result of the company's profits falling by 23% in 2019, much of it due to changing shopping habits. The 2020 coronavirus pandemic has changed many kinds of behaviour in places, which might continue in the future.

Summary

- Place is a part of everyone's life. We all give different places different meanings.
- A place's identity will be seen differently by different groups of people, and by those who seek to manage and represent the place on our behalf.
- People become attached to places at a variety of scales.
- Places adapt and change over time, and contain elements whose identities have been modified by new social and economic events.
- Places are interdependent — increasingly as a consequence of globalisation.
- Always be able to use your home places to illustrate points.

Changing places: meaning and representation

How are places given meaning and represented by people?

People view places differently as a consequence of their differing identities, perspectives and experiences. These include age, economic status, ethnicity, gender, ideology, language, politics, race, religion and social class.

- Jerusalem has a different meaning and significance according to whether you are a Jewish, Muslim or Christian person. It is a divided city, whose divisions reflect the engagement of the Palestinian and Jewish populations with the city over time (**difference**).
- In Gateshead, teenagers expressed a fear of certain areas in the town, with women (**gender**) being more fearful, despite evidence that they were less victimised than men. Visitors to a place perceive it differently from residents (**economic status**, **social class** and **language**).

Exam tip

Remember that a resident may have a perception of a place based on obsolete factors that no longer influence its character. A newcomer to the area may consider such factors unimportant and thus have a very different perception.

- In another study, Asian and Afro-Caribbean 16–24-year-olds felt safe in the terraced housing community where they lived (**race**, **ethnicity** and **age**), whereas other ethnic groups felt unsafe there, especially at night.

The meaning and representation of Chichester Harbour Area of Outstanding Natural Beauty

Chichester Harbour, an Area of Outstanding Natural Beauty (AONB), encompasses a wide range of environments and urban places within its 74 square kilometres. The Chichester Harbour Conservancy was established in law in 1971, and administers the area and produces plans for its future. The AONB (Figure 2) contains coastal villages, such as Bosham and West Wittering, and is fringed by former market towns, such as Emsworth, Havant (a suburb of Portsmouth) and the seaside resort of Hayling Island. People's perceptions of this place vary considerably, and how they identify with it will depend on how they perceive where they live and the experiences they undergo in that **locality**. There are several groups who attribute **meaning** to the AONB (Table 3).

Table 3 Groups involved with Chichester Harbour AONB

Group	Factors influencing perception
Chichester Harbour Conservancy: produces the management plan and administers the AONB. It belongs, politically, to two county areas — West Sussex and Hampshire. The main aim is to conserve the natural environment	Politics Ideology
Residents: 10,585 people live within the AONB. 31% are retired. Since 2011 there has been a 16% increase in those aged 70–79 and a 10% increase in 80–89 year olds. 30–49 year olds have decreased in number. Incomes are significantly higher than the UK average. Part of their council tax goes towards maintaining the harbour	Economic status Social class Age Gender
Second-home owners: form 25% of the housing stock. Property prices are 5% higher than average for the southeast region	Economic status Social class
Parish councils: concerned about sea-level rises, and sewage generated by the increasing number of homes	Politics
Recreational visitors and tourists: there are 1.5 million recreational and tourist visits per year. 45% of these visitors travel from under 10 miles away, mainly to sail	Economic status Social class Age
Sailing clubs and educational activity centres: there are 14 sailing clubs, 5,200 moorings/berths and 10,500 registered vessels. Boat owners pay harbour dues and mooring fees	Age Gender Economic status Social class
Organisations managing environmental conservation: these include the National Trust, Natural England and those who administer Local Nature Reserves and Sites of Special Scientific Interest (SSSIs). These organisations know that management is within the context of a 5.2 mm/year rise in mean sea level since 1991. 41% of the area is below mean high water springs	Ideology Politics

Exam tip

Factors such as age, gender, race, ideology and experience will influence your own perception of a place. It is important that you carefully consider all the evidence presented before making a judgement about a location.

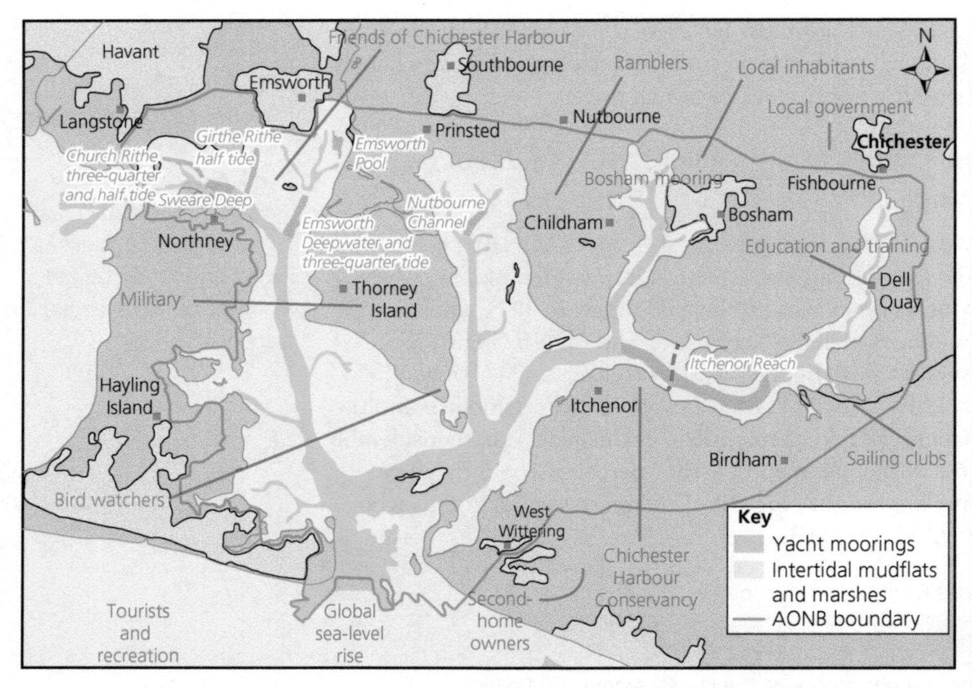

Figure 2 Chichester Harbour Area of Outstanding Natural Beauty and the pressures on the area

Knowledge check 4

For each of the groups listed in Table 4, copy and complete the columns to illustrate how people engage with the harbour as a place and perceive what is happening, and the experiences that may have influenced some members of that group in engaging with and perceiving the harbour area.

Table 4 Engagement with, and perception of, Chichester Harbour AONB

Group/ organisation	Engagement with harbour	How harbour locality is perceived	Experiences that influence perception of place
Local residents			
Second-home owners			
Yacht clubs			
Parish Council			
Conservancy			
Farmers			

Representations of place in advertising and promotional material through different media and publications

Table 5 lists a selection of groups who use the media to advertise and promote the city of Portsmouth. Individuals can also represent a place, through social media such as Facebook and Instagram. **Representation** can also be based on one's personal experience on a single occasion, such as visual impressions when house hunting, a traffic jam and certain events. A place can be represented negatively because it floods, or positively because of individual or team sporting success.

Exam tip

Make sure you understand and have examples of the different ways places are represented, including where possible, your home place.

Table 5 A selection of representations of Portsmouth

Source	Who	Why	Outcomes
City Council	The party in political power	Maintain commercial and business profile of city; portray as a welcoming city	Retail developments, new business areas, recreation and tourism income; welcoming increased numbers of refugees
Visit Portsmouth	City Council	Promote tourism with images	Tourist visits to attractions
investinportsmouth.co.uk	City Council, business groups	Loss of employment in naval dockyard	Regeneration and redevelopment; tourism
Waterfront and island city	City Council	Uniqueness of the whole place	Tourism; in-migration of jobs and people
Portsmouth FC	Owned by Tornante Company, an American-owned investment firm	Advertise club and build up the fan base	Improving position lost after 2010; team's supporters as a community
University of Portsmouth	University and its governors, Universities UK	Status in UK and overseas	League table position; staff and student recruitment; the student spend
International ferry port	City/ferry and shipping companies	Employment and trade; three sites of commercial docks	Tourism, including cruises; imports (bananas)
Cathedrals	Roman Catholic and Anglican churches	Community and spiritual involvement	Talks on ethics —community awareness
Muslim communities	Two mosques	Spiritual and community involvement	Inter- and intracommunity harmony
Portsmouth Sixth Form College	College and its governors	Role in city community; recruitment	Equal opportunity college; enhanced status in city; a learning community
Clinical Commissioning Group	24 GP surgeries	Health of population	Healthy, employable population; prestige
IBM	Major employer	Global company; invested in area for 50 years	Do not say they are in Portsmouth but their location is in North Harbour; attract employees
BAE Systems	Defence industry in naval dockyard	Loss of shipbuilding	Government investment in naval ship repairing and servicing; maintains some jobs
The National Trust	National interest group	Potential for city to be at flood risk from global warming	Saving homes and jobs in an island city
Americas Cup Consortium	BAR and Land Rover; Ben Ainsley	Investment and publicity; good location for boat racing	Major world sporting event; jobs and income to region; prestige in UK and abroad

Contrasting images of places portrayed by formal statistical, media and popular representations

Places can be represented in a variety of ways. These can be formal, such as census data and news media, or informal, produced by people outside of the formal sector institutions. Such contrasting representations can influence the perceptions of a place. Portsmouth, for example, is portrayed very differently in the following three ways:

- A search at www.ukcensusdata.com gives access to a formal statistical summary for Portsmouth. It contains details on population, age groups, dependency ratios, marital status, country of birth, ethnicity, tenure, religion and household composition.
- The information at www.portsmouth.co.uk presents a factual but very different image of Portsmouth.
- Searching the internet for 'what is Portsmouth like?' gives access to many informal representations of the area.

Knowledge check 5

Categorise the following types of representation into formal, informal factual, informal non-factual or informal opinion:

- Office for National Statistics data
- Social media
- Television drama
- Graffiti
- Local historian's account of a past event
- Literature
- Tourist organisation promotional material

Place meanings can have an effect on continuity and change for places

The ways in which individuals perceive and give meaning to a place will differ from person to person. This in turn can result in variations in the demand for continuity or changes to a place. Visitors may wish to see an area of attractive scenery and picturesque villages, such as those found in National Parks, preserved with little change, while locals may want changes such as development of the infrastructure for modern life or even affordable housing.

Place meanings can affect the lives of people

The **meaning** of places can alter with your value system. For example, the dominance of a building such as the Principality Stadium in Cardiff or the Shard in London may be viewed negatively among those with a conservative value system, whereas those with a more liberal set of values may react positively.

An individual's perception of, and attachment to, a place can change over time along with other factors in their lives, such as age and wealth. With increasing age and wealth, for example, a person could grow more attached to a location that they perceive as meeting their current needs, such as a larger home or better proximity to health care. A change in the attachment an individual has to a place can have an impact on people's lives. For example, events that change potential visitors' perception of a location can change the number of people visiting, which may in turn impact on employment and wealth in the place. This may occur after a terrorist or natural hazard event changes a person's perception about how safe a place is to visit.

Knowledge check 6

What are the factors that may influence the perception a person has of a place?

Summary

- People will perceive places differently due to a variety of factors.
- Places can be represented in a variety of ways, both formal and informal, which may influence people's perception of them.
- Frequently, interest groups will give meaning to a place that backs up their own ideas and ideology.
- The media portray places and give them identity as a result of the events and stories that they select to publish.
- Local government and companies will also attempt to portray a place in the most favourable light in order to boost the image of the place or company.
- Changes in the perception and meaning of a place can have consequences on the lives of people attached to that place.

Changes over time in the economic characteristics of places

Models of economic and employment change in places over time

The **Clark Fisher model** is a stylised way of describing the changing balance of employment over time, and has been used mainly at a national level. The model (Figure 3) distinguishes four sectors of the economy:

1. **Primary:** the part of the economy concerned with the collection and use of natural resources.
2. **Secondary:** the manufacturing or industrial sector — the part that processes resources into goods that people want.
3. **Tertiary:** the sector that enables goods to be traded, sometimes called producer services — includes wholesaling, retailing, banking, finance and insurance, transport, and entertainment, including tourism and personal services.
4. **Quaternary:** research and development, and the knowledge economy, including IT, education and the processing of information.

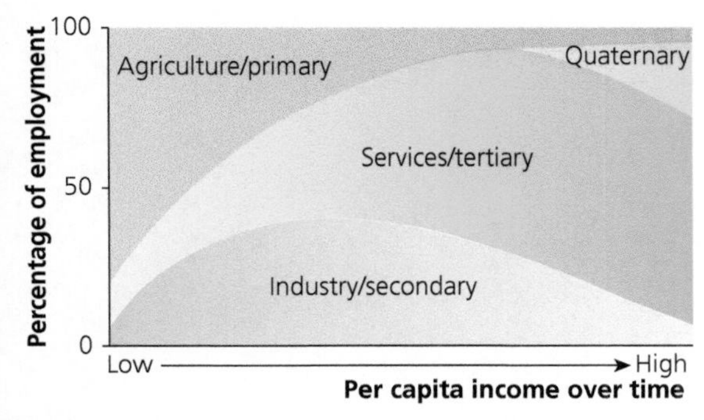

Figure 3 The modified Clark Fisher model

In 2015, the Office for National Statistics (ONS) only distinguished the following:

- Agriculture and fishing (primary)
- Services (tertiary and quaternary)
- Manufacturing (secondary)
- Construction (secondary)

Note: the bracketed terms are those used in the Clark Fisher model.

Figure 4 shows how the balance of employment has changed over the past 160 years.

Exam tip

Not all countries will follow the Clark Fisher model. The growth of a tourism industry may result in an industrial stage being missed out. Within one country different regions may have very different employment structures.

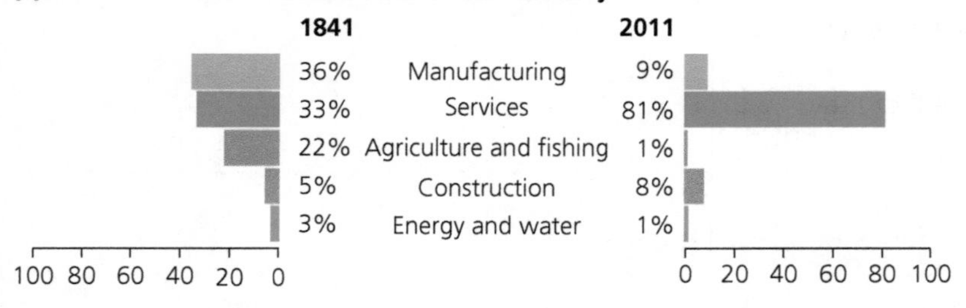

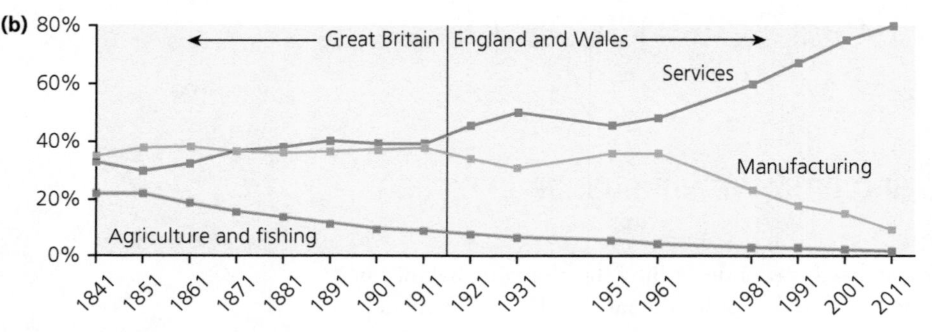

Figure 4 (a) Changes in the five employment sectors between 1841 and 2011. (b) The percentage of the workforce by employment sector between 1841 and 2011

Kondratiev waves (Figure 5 and Table 6) are another model used to describe the economic changes over time at a national level. Kondratiev waves are approximately 50 years in duration and each of the four past waves (K1–K4) has four phases: prosperity, recession, depression and recovery. Each wave is associated with the development of particular technological innovations and economic activities. The model ignores primary production. Some of these waves will have affected your home region and places within it.

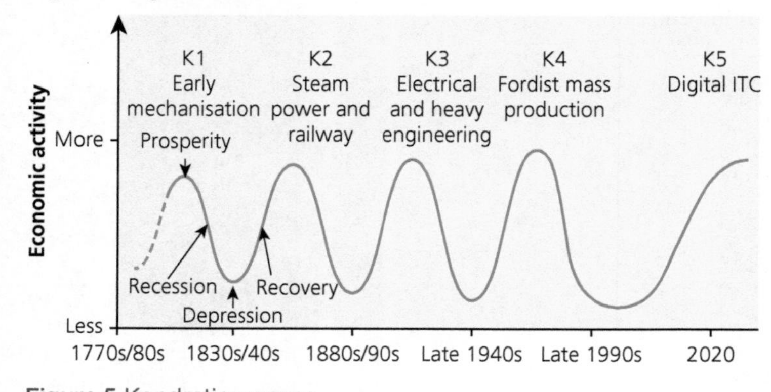

Figure 5 Kondratiev waves

Table 6 The five Kondratiev waves

Waves	1st (1770s/80s to 1830s/40s)	2nd (1830s/40s to 1880s/90s)	3rd (1880s/90s to late 1940s)	4th (late 1940s to late 1990s)	5th (late 1990s to 2020 onwards)
Main industries and/or economic activities	Water power, textiles, iron, potteries	Steam engines/ ships, iron and steel, coal mining	Electrical engineering, heavy engineering, armaments, steel ships, chemicals, dyes	Automobiles, lorries, consumer durables, synthetic materials, petrochemicals	Computers, digital technology, internet, software, optical fibres, robotics and biotechnology, universities, R&D, creative industries
Transport and communication innovations	Canals, turnpikes	Railways, steamships	Electricity supply	Highways, airports	Digital, satellites, fibre optics, wi-fi, the Cloud
UK place examples	Manchester, Bradford, Stoke	Consett, Ebbw Vale, Aberdare, Glasgow	Birmingham, Birkenhead, Port Sunlight	Dagenham, Luton, Billingham	Silicon Roundabout (Old Street), Shoreditch, Cambridge, Menai Hub
Other world places	Lille, Verviers	Ruhr cities	Ludwigshafen	Detroit, Wolfsburg, Eindhoven	Silicon Valley, Bangalore

These two models help to explain the changing dominance of activities in urban areas and, to a lesser extent, rural communities. Industrialisation is the process through which a society develops an economy based on mass and the mechanical production of goods. During the Industrial Revolution, an explosion of population and growth in urban areas located close to each other and around transport nodes, such as ports and coalfields, led to rapid urbanisation. But the advent of electricity in the third Kondratiev wave enabled manufacturing to spread beyond cities and onto cheaper land.

Exam tip

Demonstrate your knowledge of these waves when discussing the changing economy and society of places through time.

Economic change in Cardiff

Cardiff grew between 1801 and 1901 from a city of 1,870 people living within the confines of the medieval walled town to an industrial city of 164,333. The growth of steam, coal mining and iron production (second Kondratiev wave) gave rise to Cardiff becoming a major coal exporting port. The role of the wealthy Bute family in digging the canal and building the docks, and the coming of the railways between 1836 and 1853, aided the growth of the port and the city. The working population grew rapidly due to immigration. In 1851, approximately one-third of Cardiff's inhabitants came from Glamorgan and Monmouth, the surrounding counties. Others came from West Wales, Gloucestershire and Somerset. Most notable were Irish people fleeing famine in Ireland, who ended up living in overcrowded conditions in Splotlands and Newtown, described in 1855 as follows:

> **It has a low level; the houses for the most part are occupied as Irish lodging-houses and are seriously overcrowded.**

The residential areas that were developed at this time became more socially segregated. Bute Town had a high concentration of lodgers and Grangetown contained poor housing, whereas Canton, Cathays and Roath were fast-growing areas housing those who were more affluent.

A number of factors contributed to the changing economic base of nineteenth-century Cardiff.

- Population increase was rapid due to better disease control.
- Migration due to people being pushed by famine and rural poverty towards jobs and earnings, digging the docks, building homes and working in the new trading and commercial world of the city.
- The role of individuals, such as the Bute family, who used their wealth to invest and increase their own prosperity. In other places, philanthropic employers set up 'utopian' model settlements, such as Titus Salt's Saltaire, Lever's Port Sunlight, Cadbury's Bourneville and Rowntree's New Earswick.
- Transport developments enabled exports and trade to rise and made it easier to reach the city and its docks.

Competition from the docks at Barry and decreasing demand for coal resulted in the decline of Cardiff as a port, and by the 1980s large parts of the docks were derelict. Economic decline lasted until the early 1990s. Since then, several urban regeneration projects have made Cardiff a desirable location to live. Today 80% of the population are employed in the service sector, with only 6% in manufacturing. The city has a thriving financial and business services industry and, with the redevelopment of Cardiff Bay and the Millennium Centre, has seen large growth in the tourist industry with over 20 million visitors a year.

Knowledge check 7

What is meant by the term 'utopian'?

Location quotient

Location quotient (LQ) is a statistic that measures a region's industrial specialisation and concentration relative to a larger geographic unit (normally a country). An LQ is calculated as the share of an industry of the regional total for an economic statistic (number of factories, employment, etc.) divided by the industry's share of the national total for the same statistic. For example, an LQ of 1.0 means that the region and the nation are equally specialised in an activity, whereas an LQ of 1.8 means that the region has a higher concentration than the nation and an LQ of 0.5 means that it has a lower concentration. LQ is calculated using the following formula:

$$LQ = \frac{\text{\% of the total workforce in the area working in an activity}}{\text{the workforce in that activity in the country as a percentage of the total workforce}}$$

Analysing changes in LQs can highlight economic changes in places, which in turn can lead to structural changes in employment. Table 7 shows changes in the manufacture of electrical equipment in Peterborough. 2015 saw the closure of a factory manufacturing electrical goods, which was a significant employer in the area.

Exam tip

You must be prepared to calculate data and apply formulae, such as that for location quotients, in an examination. It is important that you can interpret the meaning of the result, as this may be the task you are asked to complete for the data you have been given.

If you cannot complete statistical or mathematical work in the time allowed, do not waste further time trying to answer the question, as it will likely count for no more than 5 marks at most.

Table 7 Location quotients (LQs) for electrical equipment manufacture in Peterborough 1971–2015

Year	Location quotient
1971	3.8
1979	1.7
2009	6.8
2015	2.0

External forces and factors influencing economic restructuring

A number of factors, such as changes in technology or globalisation, can result in a place changing economically with resulting changes in employment structure.

Ebbw Vale is a place that has been affected by the forces of economic change over the past 200 years (**time**). Ebbw grew as a product of the Industrial Revolution when an ironworks, powered by the coal extracted from the mines in the valley, was established in the region in the 1790s. In the 1860s it became a centre for steel manufacture and a classic case of a plant based on coal energy supplies, using locally quarried limestone in the blast furnaces and becoming increasingly reliant on imported iron ore.

In the Depression years of the 1930s there was a decline in demand for coal and only 1% of the works remained in service. The Special Areas (Development and Improvement) Act 1934 attempted to rectify economic inactivity, unemployment, poor communications, poor housing and low skill levels in the region. However, an entrepreneur, aided by government grants, built the first integrated steel mill in Europe at Ebbw Vale, which used a new American technology known as continuous hot rolling (**changing technology**). In 1947 tin plating was added to the works and upgraded in 1978 (**mitigation**). At its peak, the site supported a workforce of 16,000. However, by the 1970s the steel works generated high costs because it had to import ores from overseas and transfer them by rail to the site. It became cheaper to import steel from overseas and, as a result, Ebbw Vale was unable to find a market (**globalisation**). All steel production ceased in 1977–78 and the steelworks was demolished in 1981. Tin plating continued until 2001–02, when it was finally shut down. The resultant site/place was described as 'a 2-mile scar in the heart of a town'.

Long before the whole site was closed, Ebbw Vale bid to be a National Garden Festival site in 1992. The former iron and steel works site housed the Festival for 6 months. Since then, 'The Works' (Figure 6) has been redeveloped for housing, education (a school for 3–16-year-olds and Coleg Gwent for post-16 education), retail (Festival Park) and a hospital (**government strategies**), with £350m investment aid from the EU. In 2010 the cooling ponds were converted to a wetland centre (**environmental sustainability**). A museum has been opened in the old works offices.

The Works is a 6.2 ha site within the 38 ha Enterprise Zone in the modern administrative area of Blaenau, Gwent (Figure 6). Enterprise Zone status (page 36) enables the designated area to attract capital allowances from the government and funds for SMEs (small- and medium-size enterprises) from Finance Wales. The master plan is an 'improvement from within' approach, i.e. if you provide space and improve education and skills, business will locate here and the place will improve (**mitigation**). However, this ignores global economic development conditions, which have not been favourable since the 1990s. In 2015 Ebbw Vale and the Heads of the Valleys towns (e.g. Merthyr Tydfil and Tredegar) continued to be characterised by high concentrations of social deprivation and economic inactivity, poor health, low levels of attainment (40% unemployed or unavailable for work in 2016) and skills, and depopulation (**lifestyles**, **risk**).

In 2019 the Garden Festival site was bought by London-based investor GWM Capital, with plans to transform the area into an activity and retail hub for the wider region, which would become a major shopping destination and tourist attraction.

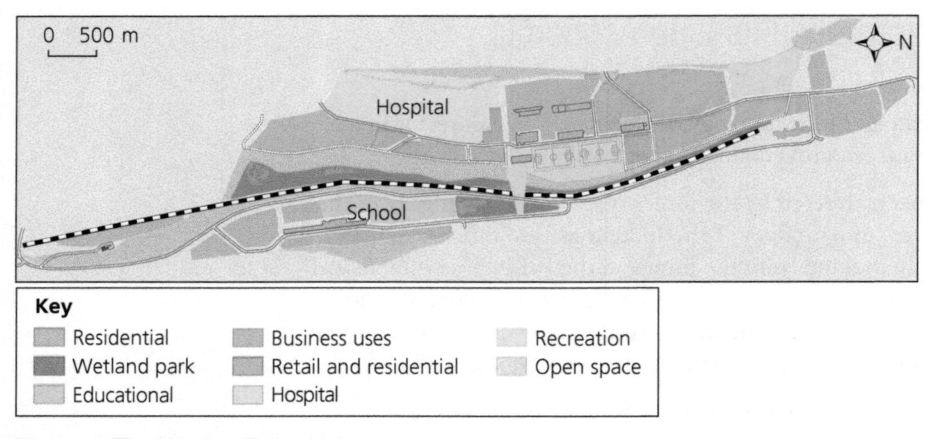

Figure 6 The Works, Ebbw Vale

A number of different factors have brought about the economic changes outlined for this place (**mitigation**):

- **Technological changes:** changes in iron and steel making, tin plating and transport technologies introducing new technologies and pharmaceuticals.
- **Government strategies:** at local (Blaenau, Gwent), regional/national (Wales/UK) and EU/international level (Special Areas Act 1934, Development Areas, Enterprise Zones, Garden Festivals, ERDF Structural Funds, creating new education facilities). Any EU funding ceases in December 2020.
- **Resource depletion:** original iron sources depleted; coal supplies costly; cheaper imports to coastal steel works (sustainability).
- **Economies of scale:** unable to compete with larger, often coastal plants, in both the UK and abroad, therefore high-cost, uncompetitive products.
- **Globalisation:** competition from overseas makers; purchase of equipment by other, larger global producers; opening of biotech company Penn Pharma in Tredegar.
- **High labour costs:** compared with global competitors.
- **Lifestyle changes:** emphasis on education and skills in the twenty-first century; environmental protection and sustainability in the new wetlands; the addition of an environmental resource centre.
- **Boosterism:** publicity to create an image of modernity represented by new activities and yet keeping a sense of past Ebbw in the museum — a listed building.

Exam tip

Always have examples to support every point you make. They can be extended examples for use in essays or brief examples to support a set of reasons or factors.

The decline in primary employment in rural areas and in secondary employment in urban places

The employment structure of the UK has changed significantly since the 1920s (Table 8).

Table 8 UK employment structure 1920–2016

Year	Primary (%)	Secondary (%)	Tertiary (%)
1920	14	34	52
1960	7	40	53
2016	1.3	15	83.7

Rural areas have experienced changes in the employment structure due to the decline in primary activities. Causes including the impact of mechanisation, resource depletion and globalisation have all resulted in job losses. Some primary jobs are seen as physically demanding and are less attractive to younger workers. The average age of a farmer in the UK is 59, and is rising because older farmers tend not to retire.

The challenges are great for primary industries, especially in the more remote areas. In Wales, 60% of its 20,000 km^2 is agricultural and 9.5% is forested. The population densities are low, the average farm size is small (59 ha) and accessibility is by narrow roads. ICT and broadband connectivity are poor. It is difficult to provide services when there are few settlements that can be hubs to serve a wide area, all of which has an impact on the lives of people living there. Farms may lack a skilled workforce, so much of the Rural Development Planning money is being invested in up-skilling 13,000 farmers. Out-migration of the young in order to gain qualifications elsewhere leaves a young population with low-level qualifications and high unemployment; 6.9% in rural areas are unemployed (**risk**).

These changes can have both negative and positive impacts on the lives of people in rural areas (Table 9).

Table 9 Impacts of the decline of primary industry in rural areas

Negative impacts	Positive impacts
■ Decrease in employment opportunities ■ Changes in rural communities due to rural-to-urban migration in search of jobs ■ Ageing population remains in the area, reducing community diversity. They may be disadvantaged by the costs of accessing services. The ability to continue working in activities such as farming may decrease ■ Negative multiplier effect results in declining services ■ Environmental concerns over industrial dereliction in rural areas (e.g. South Wales) ■ Increasing isolation (especially in farming) ■ New jobs on short-term or zero-hours contracts	■ Reduction in environmental pollution ■ Potential new employment opportunities due to industrial heritage tourism ■ Removal of old industrial buildings creating a better-quality environment to live in ■ Potential new leisure and recreation facilities (e.g. flooded gravel pits) ■ Land available for new housing, easing the housing shortage ■ Importation of cheaper goods allows disposable income to go further, improving welfare ■ New tertiary jobs may be less physically demanding and located in healthier environments

Urban areas have experienced changes in the employment structure due to the decline in secondary activities. In 1980 the headline 'The murder of a town' was used to highlight the loss of 3,700 jobs in Consett when the steel works was closed, resulting in 36% unemployment. The following section provides further case studies illustrating the impact of decline on urban places.

Deindustrialisation is the decline in manufacturing industry. It can be caused by:

- a fall in the output of manufacturing
- the development of new, improved products
- the growth of cheaper imports of the same products
- a drop in the number and share of employees in manufacturing as a result of factory closures, automation and lay-offs

Not every industrial city has been able to remain successful. Detroit lost 58% of its population between 1950 and 2008 due to deindustrialisation, automation and foreign competition for the automobile industry. Declining population due to a downturn in a city's port and shipbuilding industrial base has affected Liverpool and Glasgow.

Cheaper foreign competition from South Korea and Japan killed shipbuilding on the rivers Tyne and Wear. The loss of cotton textile and mining industries has particularly affected parts of Bolton, Rochdale, Oldham, Tameside and Wigan in Greater Manchester (Figure 7). In these places, 30% of neighbourhoods are counted among the 20% most deprived neighbourhoods in England. By 2011, only 9% of the UK workforce was employed in manufacturing.

Figure 7 Employment and deprivation in Greater Manchester

Deprivation is the theme of the following section in this Guide.

Deindustrialisation affects places differently and results in different outcomes. Cities such as Stoke-on-Trent, Hull, Barnsley, Middlesbrough, Bolton and Blackburn have all been identified as struggling because new jobs are not replacing lost jobs. The qualified leave and the unqualified remain, which further reduces the attractiveness of the workforce to employers. Although many deindustrialised urban areas are deprived, they are surrounded by affluent rural areas. Hull is the ninth most deprived city in the UK, yet the surrounding East Riding is 208th.

Summary

- The Clark Fisher model provides a basic explanation of the changing balance of employment in developed economies over time.
- Kondratiev waves help to explain the relationships between technological change and the economic development of places and regions.
- Location quotients (LQs) measure the concentration of an activity in areas/regions.
- Changes in primary employment in rural areas can have a significant impact on rural populations.
- The decline of the manufacturing industry in the UK and other developed economies is a consequence of factors emanating from both beyond the country and within the country/region.
- Deindustrialisation has been a significant outcome of the decline in manufacturing and has had a variable impact on places. It has given rise to inequalities both within and between towns and cities.

Economic change and social inequalities in deindustrialised urban places

Consequences of the loss of traditional industries in urban areas

In 2016, Tata Steel, the Indian-owned steel maker, announced 1,050 redundancies in the UK. The company employed 6,000 in Wales (4,000 in Port Talbot, where 1 in 4 people worked in the steel industry) and 750 of these jobs were declared redundant. Other Tata sites are in Llanwern (Newport), Llanelli, Shotton and Trostre. It has been estimated that the wages of Tata employees bring £200m to the Welsh economy. What are the effects of such announcements on the surrounding places?

Table 10 shows the broad consequences of deindustrialisation for selected cities that had the lowest population growth between 2004 and 2014. Three Welsh cities have been added for comparative purposes. Each of the English places illustrates direct and indirect effects on the economies and society. On every measure these former industrial cities fall well below the national average. Their contributions to the economy, attractiveness to new business, innovation as measured by patents and employment rates are below UK averages. More residents have no qualifications and fewer have high-level qualifications. Only broadband connectivity is better than the national average.

Key to Table 10	
Low-wage, high-welfare cities	Low-wage, low-welfare cities
High-wage, low-welfare cities	UK average

Table 10 Selected characteristics of deindustrialised places with the lowest population change (after Centre for Cities, 2019)

Place	Population growth 2016–17 (%)	GVA* per person (£) 2017	% residents with high-level skills (NVQ4 and higher) 2017	% residents with no qualification 2017	New business start-ups per 10,000 population 2017	Patents per 10,000 population 2017	% employment rate 2018	% households with ultra-fast broadband 2018
Blackburn	0.2	44,900	28.8	12.1	58.5	8.4	64.1	72.4
Stoke	0.3	43,500	24.5	10.8	31.5	8.8	72.9	72.2
Hull	0.2	41,400	26.7	9.8	34.3	7.3	69.7	88.0
Middlesbrough	0.1	48,000	31.0	12.0	35.0	10.7	67.8	91.0
Burnley	0.2	48,700	27.8	9.2	36.1	6.6	71.5	45.7
Sunderland	0.0	53,100	27.3	8.7	31.0	6.9	71.0	54.8
Newport	0.9	45,900	34.6	7.6	79.8	12.1	74.2	56.7
Swansea	0.4	44,300	31.8	10.1	31.5	10.5	67.7	71.0
Cardiff	0.7	51,000	48.0	6.6	45.4	24.8	70.6	83.6
Lowest achieving	Luton −0.6	Hull 41,400	Mansfield 17.8	Belfast 16.1	Plymouth 30.6	Wigan 3.1	Blackburn 64.1	Aberdeen 2.3
Highest achieving	Coventry 2.0	Slough 82,000	Oxford 63.0	Exeter 2.7	London 101.1	Cambridge 269.8	Worthing 85.8	Luton 95.0
UK	0.6	57,600	38.4	8.0	57.8	3.6	74.9	56.1

*GVA: gross value added

Wages and welfare payments provide a further indicator. All but Cardiff are what are classified as 'low-wage, high-welfare' cities. Those in the lowest achieving row are all 'low-wage, high welfare' cities with the exception of Luton and Aberdeen. In contrast, the best achieving cities are all 'low-wage' cities with the exception of Worthing, although not all are 'high-wage' cities.

Figure 8 maps cities with the highest percentage of high earners and the highest percentage of low earners in the workforce. Those places with low earners are mainly the cities of nineteenth-century industrial regions of England and Wales, whereas the high-earning cities are associated with employment diversity relating to the growth of the tertiary and quaternary sectors (**causality**, **inequality** and **difference**).

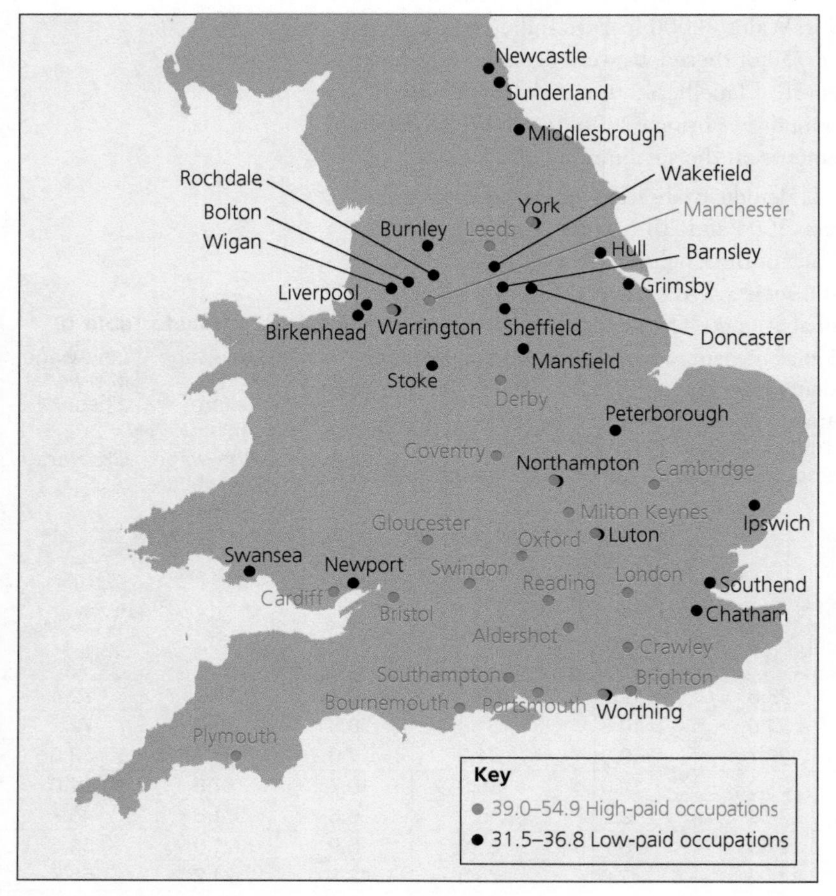

Figure 8 Inequalities in city occupations 2011

Deprivation

One social outcome of deindustrialisation is **multiple deprivation**. The UK government has developed a multiple deprivation index based on seven quality-of-life indicators: income, employment, health, education, access to housing and services, crime rate and living environment. These problems tend to be more visible in urban areas, but they can also affect rural areas.

Exam tip

If you live in or near a city you can research its characteristics at the Centre for Cities website: www.centreforcities.org.

Knowledge check 8

What is meant by 'gross value added'?

Knowledge check 9

Some places in Figure 8 occur in both high- and low-paid categories, for example Worthing, Luton and York. Why might this be so?

Exam tip

The indicator 'living environment' refers to the conditions in which a person is living, such as quality of housing or air quality.

Exam tip

When using a case study example of deprivation, try to link it to the indicators in the index rather than generalising.

Deindustrialisation can impact on quality of life in many ways. Limited job opportunities can result in a lack of income due to unemployment or new jobs being lower paid. The lack of income limits access to housing, which has become unaffordable. Services may close due to the lack of spending power of the population. Poorer living conditions combined with stress-related illnesses can cause declining health. Areas of deprivation may have reduced access, or a negative attitude, to education and often suffer higher crime rates.

Unemployment resulting from deindustrialisation can therefore result in a cycle of deprivation (Figure 9).

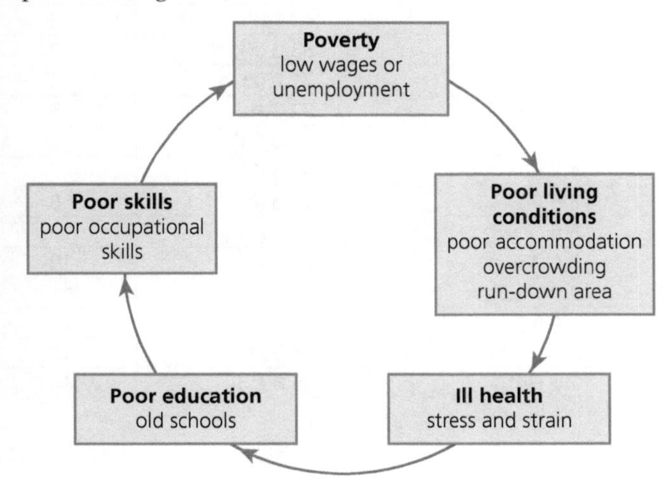

Figure 9 The cycle of deprivation

Figure 10 maps the areas of greatest and least deprivation in the city of Leicester. Similar maps can be found for almost every unitary authority in the country, so that you can examine and understand the nature of the deprived localities in your own place.

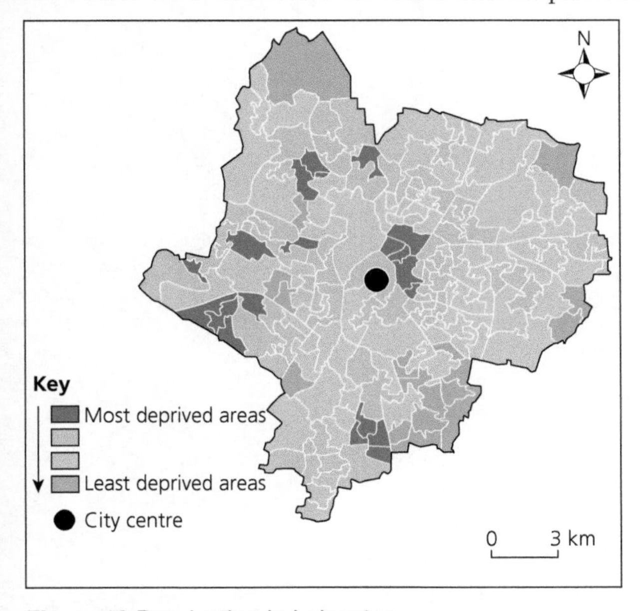

Figure 10 Deprivation in Leicester

One measure used to compile the education domain in the index of deprivation is the percentage of the population with no qualifications. Table 11 shows data for the 14 wards of Portsmouth and their ranking on the index of multiple deprivation (IMD) for the city (**inequality**). Figure 11 shows the location of the wards in the city.

Table 11 Percentage of population over 16 with no educational qualifications, % overcrowded households, % terraced housing and IMD rank for the wards of Portsmouth

Ward name	% no qualifications (rank)	% overcrowded households (rank)	% terraced accommodation (rank)	IMD rank (Figure 11)
Charles Dickens	31.5 (2)	8.8 (1)	14.2 (14)	1
Paulsgrove	32.8 (1)	5.0 (7)	42.1 (7)	2
Nelson	27.8 (3)	6.1 (4)	63.3 (5)	3
Fratton	21.7 (6)	6.2 (3)	71.7 (2)	4
St Thomas	15.9 (12)	7.9 (2)	22.2 (12)	5
St Jude	13.3 (13)	5.6 (5)	23.0 (11)	6
Cosham	25.1 (4)	4.2 (8)	38.6 (9)	7
Hilsea	23.2 (5)	3.8 (9)	40.5 (8)	8
Milton	19.4 (8)	3.5 (10)	70.7 (3)	9
Baffins	21.5 (7)	3.1 (13)	61.9 (6)	10
Eastney and Craneswater	16.8 (11)	3.2 (11.5)	38.4 (10)	11
Central Southsea	11.6 (14)	5.4 (6)	72.0 (1)	12
Copnor	19.2 (9)	3.2 (11.5)	66.3 (4)	13
Drayton and Farlington	17.9 (10)	1.5 (14)	18.8 (13)	14

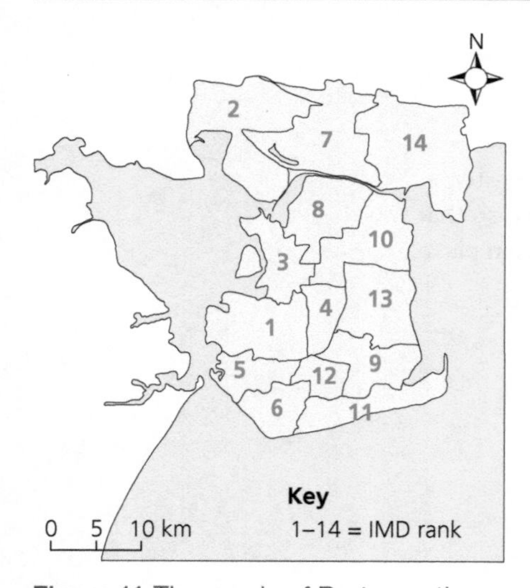

Figure 11 The wards of Portsmouth

Knowledge check 10

Perform a Spearman rank test (R_s) using the data for % no qualifications and the IMD rank in Table 11.

A Spearman rank test (R_s) using the data for overcrowding and IMD rank gave a result of 0.81. A test using % terraced accommodation and IMD rank gave a result of –0.18.

What do these results indicate?

The **Welsh index of multiple deprivation (WIMD)** has identified 389 of the most deprived areas in Wales. The components of the index are shown in Figure 12. Each domain contains further indicators — those for access to services are given. In many rural areas, the threshold for these services is not reached.

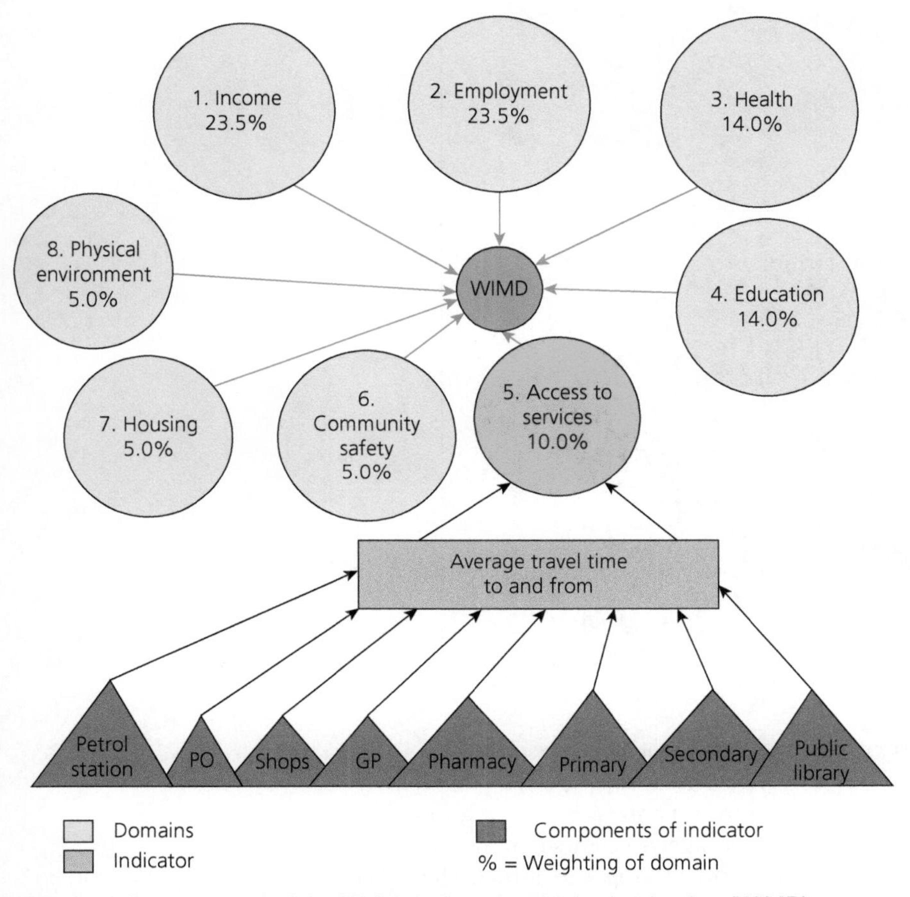

Figure 12 Components of the Welsh index of multiple deprivation (WIMD)

Figure 13 maps deprivation in Wales in 2014. The areas of highest deprivation are mostly found around the traditional industrial areas of South Wales. There are some deprivation hot spots in rural Wales. They tend to be small towns (Newtown), the most remote mountainous areas and the extreme periphery (Llyn Peninsula). Deprivation is concentrated in urban areas, with only 2% of the most deprived areas in Wales being rural. However, 18% of the most deprived live in rural areas. These data are based on extensive statistical areas, but deprivation is not necessarily concentrated, so local and regional government can miss it because individual cases get overlooked.

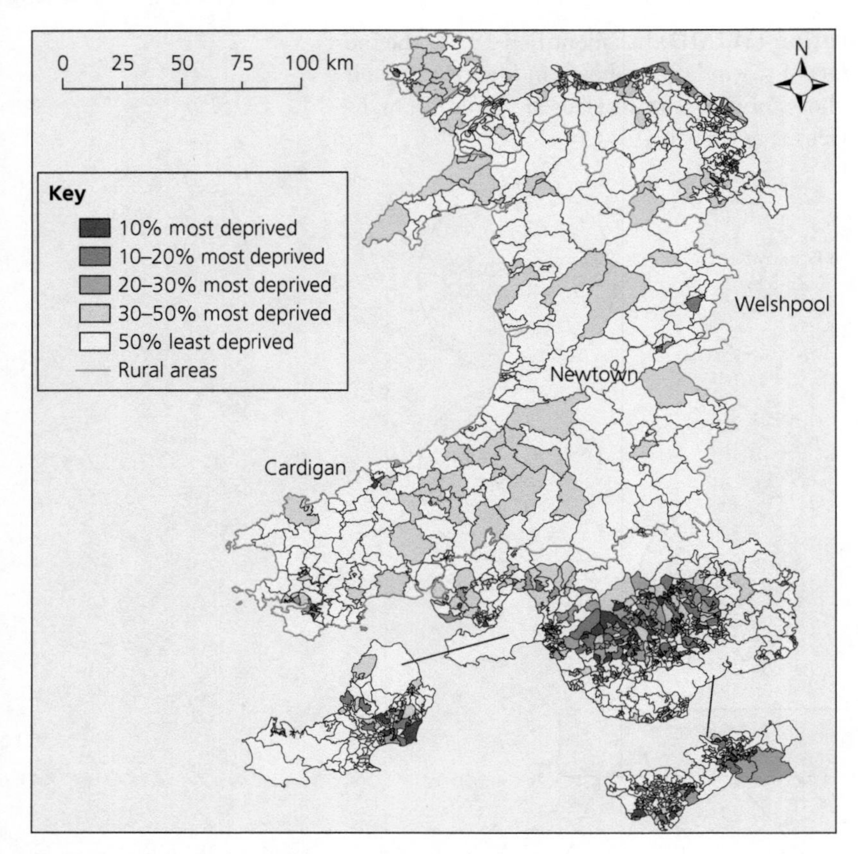

Figure 13 The Welsh index of multiple deprivation

Many areas of deprivation suffer from territorial stigmatisation, where an area is typecast because of its reputation and perceived changes to the population mix of its neighbourhoods (**identity**). In some cases, these areas may be perceived as 'ghettos' due to the ethnic mix of the people living there. The term 'ghetto' comes from Renaissance Venice, where the Jewish population was forced to live in a separate district. Thus, ghettoes enable the dominant social groups to isolate a community in a distinct, identifiable area.

In Europe, it has been partly a consequence of deindustrialisation that some of these areas have been described as 'decomposing steadily'. St Pauls, Bristol is a case in point; close to the city centre, partly destroyed by bombing and rebuilt with social housing, together with some older squares and terraces, it became an area of Afro-Caribbean immigrants intermixed with lower-paid and unemployed people in the 1950s and 60s.

The area became stigmatised because it was associated with riots in 1980, drug gangs in the early 2000s and the development of an alternative culture (**causality**). Because of this stigmatisation, some taxi drivers will still avoid parts of the St Pauls area at night. These views have influenced efforts to instigate neighbourhood renewal. Parts of the area are undergoing gentrification (page 44), which appears to focus on physical infrastructure rather than issues of employment and inequality. Some original locals are resentful because they are excluded by being priced out of housing and the new services. The council has suggested renaming the Malcolm X Community Centre, set up after the riots, to something that reflects the changes in the population, an idea that has caused resentment in some of the population. In 2018 posters appeared in St Pauls supporting the return of the area to its old, poorer environment, to ease social exclusion.

Social exclusion

Social exclusion involves being outside of, or marginalised from, mainstream society, its resources and the opportunities provided by them. It can result from the impact of deindustrialisation and the resultant deprivation. It often involves **stigma**, or severe disapproval of an individual or group. It can be multidimensional — a result of class, gender identity, race, ethnicity, sexual orientation and/or age (**difference**). Homeless people are an excluded group, some of whom may be in the predicament due to loss of employment. They are then viewed as unproductive and disaffiliated. They are part of the **cycle of deprivation** (page 29), a sequence of events that disadvantaged people experience in which one problem, such as lack of work, leads to other problems and so makes things worse.

Socially excluded individuals tend to drift towards the city centres, which are seen to have the best environments and public spaces for their lifestyles. The reaction is to 'design out' the problematic activities and introduce cultures of respect. Extreme examples include 'hostile architecture' to prevent homeless sleepers, such as slanted benches, benches with arm rests along their length, spikes in paving areas and boulders on land under bridges. Exclusion may not reach these levels, but elements of the process can be seen in:

- stigmatisation of the people living on council estates with high unemployment, often negatively referred to as 'shirkers' or 'benefits cheats'
- neighbourhoods regarded as 'no-go areas' because of petty criminality, gang activity and mental health issues
- the development of red-light districts
- costs of rental housing and the council house right-to-buy policy

A consequence of exclusion is the creation of socioeconomically defined areas in settlements depending on people's ability to pay, earning power, social status and, in the past, zoning (**causality**). Areas where excluded groups cluster can go into decline due to the lack of funds to maintain or improve the area. This in turn may increase the level of exclusion.

Pollution levels and deindustrialisation

In 2013 fuel burning accounted for 83% of pollution. Closure of heavy industries such as steel works and coal-fired power stations has reduced the amount of sulfur dioxide (SO_2) in the air. SO_2 emissions declined by 97% between 1970 and 2016, and nitrogen oxide by 72%. Between 1970 and 2013 industrial combustion emissions fell by 94%. Most of the decrease took place between 1970 and 1985 with the decline in energy-intensive iron and steel, and other heavy industries. The closure of the Redcar steelworks has helped reduce the UK's carbon emissions since 2015. There has been also been a decline in the use of coal and fuel oil in favour of natural gas in power generation. To meet the Gothenburg Protocol target in 2020, the UK will have to further cut emissions by 26%. If you are interested in the effects of the many pollutants on people, go to www.naei.defra.gov.uk/overview/pollutants?pollutant_id=8.

The reduction in older industries in inner city areas can result in reducing atmospheric pollution levels due to a decrease in traffic congestion, as well as a reduction in noise pollution from the old manufacturing processes. Where regeneration of an area occurs, it is possible to remove or improve old industrial sites which may have been considered as visual pollution as well as reducing the amount of contaminated land.

Exam tip

Many of the factors resulting in social exclusion in urban areas also apply to rural areas, although the spatial segregation of excluded groups may not be present.

Government policies in deindustrialised places

This is an ever-changing topic, because governments at all levels, from international to local, continually alter policies. As a result of leaving the EU, many of the following funds will not be available to the UK after 2020.

International European Structural and Investment Funds 2014–20

European regional policy is delivered through three main funds: the European Regional Development Fund (ERDF), the Cohesion Fund (CF) and the European Social Fund (ESF). Together with the European Agricultural Fund for Rural Development (EAFRD) and the European Maritime and Fisheries Fund (EMFF), they make up the European Structural and Investment (ESI) Funds. All of these programmes are now under one umbrella: the **Growth Programme**.

Exam tip

Make sure that you are able to describe the distribution of features or regional variations shown on a map. Do not make silly errors, such as referring to the top or bottom of an area. Use compass directions. You may need to use the scale to describe the extent of a feature.

The European Regional Development Fund

The European Regional Development Fund (ERDF) commenced operation in 1975. Its main objective is to support projects and activities that reduce the economic disparity within the member states of the EU. It financially aids projects that stimulate economic development and increase employment by supporting inward investment and retraining in the poorest regions. It helps to preserve natural environments in order to improve the quality of life of residents and make regions more attractive to tourists and investors. Infrastructural investments and educational retraining schemes are supported, and there is help to promote regional development and reduce the gap between the wealthiest and the poorest regions (**adaptation** and **mitigation**). Figure 14 shows assisted areas in the UK.

The priorities are as follows:

- Strengthening research, technological development and innovation (tertiary growth).
- Enhancing information, communication and technology (tertiary and quaternary growth).
- Enhancing the competitiveness of small- and medium-sized enterprises.
- Supporting the shift towards a low-carbon economy.
- Promoting climate change adaptation, risk prevention and management.
- Preserving and protecting the environment and promoting resource efficiency.
- Promoting sustainable transport and removing bottlenecks in key network infrastructures.
- Promoting social inclusion, and combating poverty and any discrimination.

Exam tip

The present-day economic climate means that policies are continually changing and evolving. Leaving the EU means that some funds will no longer be available. It is important to keep up to date with events and policies that will influence your home place and any areas you have studied.

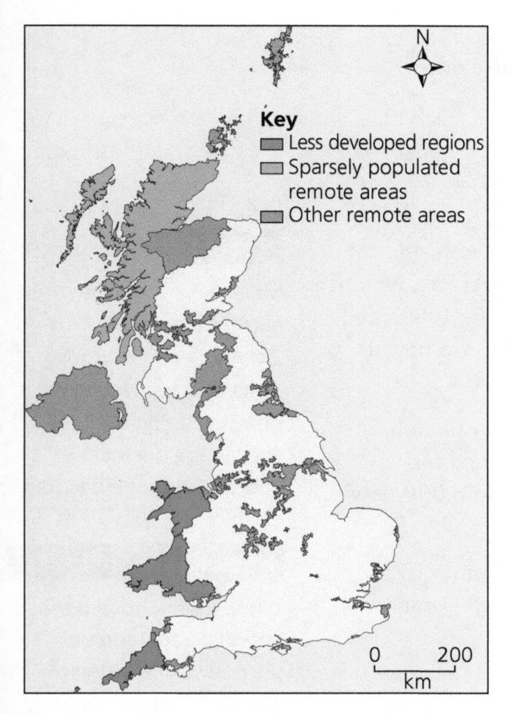

Figure 14 The assisted areas in the UK

There are three types of regional support in the UK:

- **More developed areas.** These cover most of England and are designed to reduce economic, environmental and social problems in urban areas, with a special focus on sustainable urban development.
- **Areas that are naturally disadvantaged by remoteness/mountains and are sparsely populated.** The most peripheral areas also benefit from specific assistance from the ERDF to address possible disadvantages due to their remoteness, such as the Scottish Highlands and Islands.
- **Less developed regions.** Only one region in England falls into this category, Cornwall and the Isles of Scilly, while the Welsh regions are West Wales and the Valleys.

The European Social Fund

The European Social Fund (ESF) aimed to:

- tackle poverty and social exclusion by increasing employment and helping people to access sustainable employment
- invest in skills and improve the diversity of the workforce
- invest in young people with the necessary skills for a challenging, knowledge-based economy

All of these are examples of **mitigation**.

To be successful the programme aims to:

- reduce poverty
- increase the skills levels of the workforce, and reduce the number of people with no skills or basic skills

- increase youth employment and attainment
- reduce inequalities in the labour market among women and recognise other disadvantaged groups

All these involve **adaptation**, **risk** and **mitigation**.

National funding: Enterprise Zones

The creation of new **Enterprise Zones** (**EZ**s) in 2016 means there are now 48 located in England and seven in Wales. Scotland has four **Enterprise Areas**, which are spread over 15 locations. EZs were successful when established in the 1980s and were reintroduced after 2012 as part of long-term economic plans. They are now a part of the **Local Enterprise Partnership (LEP)** programme.

Businesses locating in an EZ can benefit from business rate discounts or tax relief, simplified local authority planning regulations and infrastructure designed for businesses. In England the EZs have attracted 877 new businesses and 38,000 new jobs, with £3.5 billion invested from the private sector.

Figure 15 shows the impact on employment in Swansea in 2013. The former EZ Swansea 1 was the location for 25% of new private-sector jobs in the city — more than in the city centre (18%).

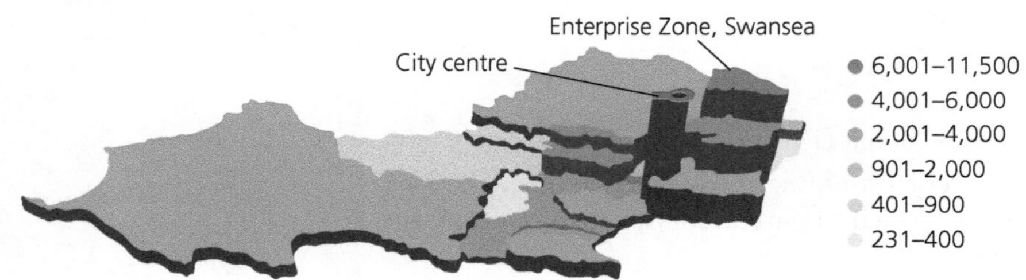

Figure 15 Employment change in Swansea 2013

Exam tip

Make sure you have current examples. Research an Enterprise Zone (EZ) to investigate how it is trying to encourage businesses and the impact it has had.

Knowledge check 11

What are the advantages and disadvantages of portraying the data in Figure 15 in this way?

Exam tip

You must be able to evaluate the merits and drawbacks of every form of data portrayal geographers use.

Local Enterprise Partnerships

Local Enterprise Partnerships (LEPs) were established in England in 2011, forming a voluntary partnership between local authorities and businesses to promote economic growth and job creation in local areas.

Solent LEP (Figure 16) has focused on six major areas within the LEP, one of which, Welbourne, was a greenfield site. This area was awarded Garden Village status in January 2017. Over the next 20 years the LEP is planned to have 6,000 new mixed-tenure homes, providing housing needs for over 13,000 people and creating around 5,700 jobs. Solent LEP investment has been over £130m, with £13m invested in over 270 small- and medium-size businesses, securing over 1,500 of the 3,000 jobs that have been created. Over 2,500 new homes have been completed and 7,200 learners have been assisted.

Community grants are being used to assist wards with high levels of unemployment and poor social cohesion. There are 16 wards on the Isle of Wight and four wards in Havant, all of which are located on former council overspill estates built in the 1950s, which are eligible for funding.

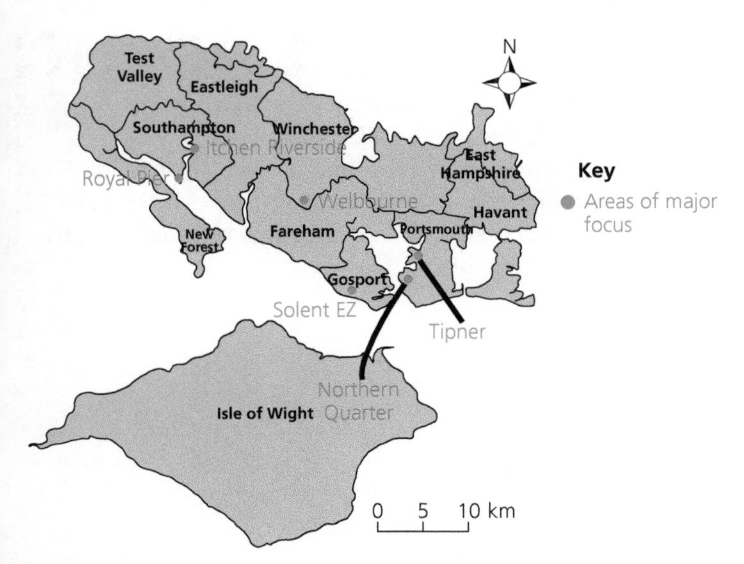

Figure 16 The Solent LEP, Solent EZ and its six brownfield regeneration sites

Foreign Direct Investment

Foreign Direct Investment (FDI) is an investment made by a company based in one country, into a company based in another country. Most FDI takes the form of investment by large multinational companies (MNCs), often based in the USA, EU, Japan, China and India. Economies with skilled workforces and good growth prospects tend to attract larger amounts of FDI. In 2017 FDI into the UK was £1,336.5 billion. In 2017–18 there were 2,072 FDI projects in the UK, creating almost 76,000 jobs and safeguarding a further 15,000:

FDI can take the form of major MNC investments, such as Toyota at Burleston, Nissan at Sunderland, Lockheed Martin Space in Harwell and BMW/Rolls Royce most recently in Bognor (**interdependence**). It can also occur in the built environment, such as the Shard, Malaysian investment in the Battersea redevelopment and UAE interests in the regeneration of East Manchester (Etihad Stadium). Not all schemes are new because some FDI comes through the takeover of existing companies. Masdar from the UAE is now the largest wind-farm operator in the UK. Hong Kong investors now own the port of Felixstowe.

FDI also attracts tertiary and quaternary industries (pages 38 and 50). Cray Computers (Seattle) opened its first office outside the USA in Bristol. A third of FDI projects in 2015 were in the quaternary sector.

Investments have often been associated with government policies to regenerate places or areas, as is the case with Nissan in the North East. The Welsh government assisted in the expansion of R&D facilities for an Israeli IT company in Newport. However, many MNCs still prefer to invest in successful places (**identity**), and not all FDI targets deindustrialised areas.

MNCs involved in FDI aim to gain some benefit from their investment, such as access to foreign markets or avoidance of trade barriers. Political events may result in changes to the amount coming into a country if the investment seems less favourable.

Exam tip

With all the economic changes that are taking place, it is important to have examples of FDI that are up to date, and to understand the impact the FDI has had in an area.

Summary

- Deindustrialisation affects the economic diversity of places, which has a range of social effects, especially deprivation.
- Multiple deprivation will vary between and within places/regions. It can be statistically analysed using secondary sources from the census.
- Rural areas have also lost employment in traditional primary industries.
- Governments at the international, national and local levels have policies to mitigate the loss of employment and to create opportunities for new employment.

The service economy (tertiary) and its social and economic impacts

In 2018 83% of the working population of the UK was employed in the service sector (including quaternary — page 50), compared with 76% in 2001. 92% of working women are employed in the sector and 71% of men. The service and quaternary economies are the main employers in UK settlements, especially London, where 94% of workers are employed in the tertiary sector.

In 1970 only 55% of the UK's GDP came from the service sector. This rose steadily during the 1980s and 90s to reach 73% in 2001. Today 79% of the UK's GDP comes from the service sector. When the tertiary sector comprises the biggest element of the economy it is known as **tertiarisation**.

Factors that promote service sector growth

The tertiary or service sector grew to support the manufacturing sector, as industrialists needed to finance growth, buy raw materials and market their products. It also grew as workers became more prosperous and they, like the companies they worked for, required banks, insurance and lawyers. In order to ensure maximum contact with their clients and rapid communication of ideas these services were initially located in city centres. Government also grew at both national and local levels, often leading to the construction of large town halls and administrative districts that reflected the wealth of the city and its region (e.g. Cathays Park, Cardiff) (**causality** and **time**). Firms that sell knowledge, such as finance, law and marketing, benefit from proximity because they can gain from the exchange of knowledge when they are clustered.

Rising affluence

Prosperity and affluence have also encouraged the growth of the service sector; people want to invest and bank their wealth, insure their property and possessions, purchase goods and have improved leisure time. More disposable income as a result of higher wages and a mortgage-free, ageing population, has led to the growth of leisure industries (**recreation** and **tourism**). The increased prosperity combined with changes in tastes has led to the growth of certain retail services, such as coffee shops.

Technological change

Technological inventions have enhanced the clustering of office buildings in central business districts (CBDs) and city centres. The development of steel-framed skyscrapers has enabled vertical development, while transport developments — especially the building of tramways and, in the larger cities, underground lines — have enabled the growing labour force to commute to work over greater distances (**causality**).

Inventions and developments in communications have enabled quicker communication over greater distances, allowing businesses to provides services further afield.

Changing transport technologies have also enabled retailing, offices and leisure industries to disperse beyond the city centre, across urban areas. Rail travel and mandatory annual holidays gave rise to the growth of seaside resorts in the nineteenth century. Subsequent developments in air travel have led to overseas package holidays and increasingly long-distance tourism and ecotourism (**causality**, **difference**, **place**).

Changing communication technologies are also altering the distribution of the service economy. Online booking has largely replaced travel agencies. Hotels and conference centres have become key functions for business life and tourism in most major cities, for example Bournemouth International Centre (BIC) and Venue Cymru in Llandudno.

Retailing, commercial and entertainment expansion in some central areas

UK city centres are the location of 72% of all highly skilled jobs and are 21% more productive than non-urban areas (Figure 17). Graduates, attracted to the concentration of knowledge-based jobs, take half of the jobs in city centres. A central business district (CBD) is a more confined area, associated with offices, administration and retail, whereas the central area includes other land uses, such as residential and leisure. Nearly all large cities have seen job growth, although many medium-sized urban areas have not.

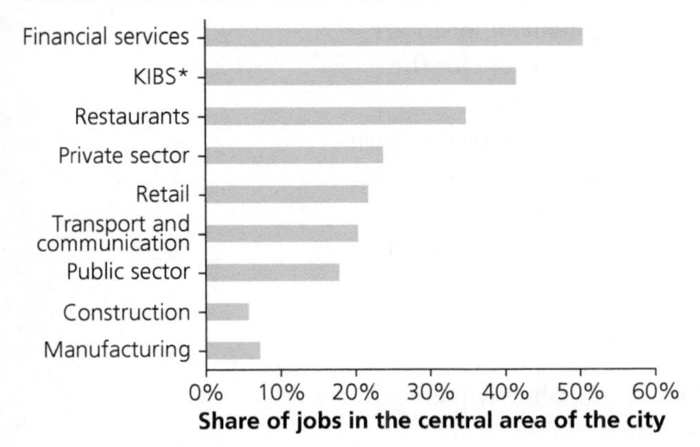

Figure 17 The concentration of jobs in central areas of cities in 2011

Benefits of city centre locations

- **Agglomeration** and **proximity** are the prime benefits of a city centre location for business and retailing (Figure 18). The centre shares the infrastructure (roads, railways) of the entire city. The urban area has a large pool of workers that employers can tap into. The centre is the area where ideas and information can be sold and exchanged, much by face-to-face contact, which is the traditional method and is still important today (e.g. in clubs such as Soho House in London). This is known as **knowledge spillover**. Finance, law and marketing are major gainers in many city centres.

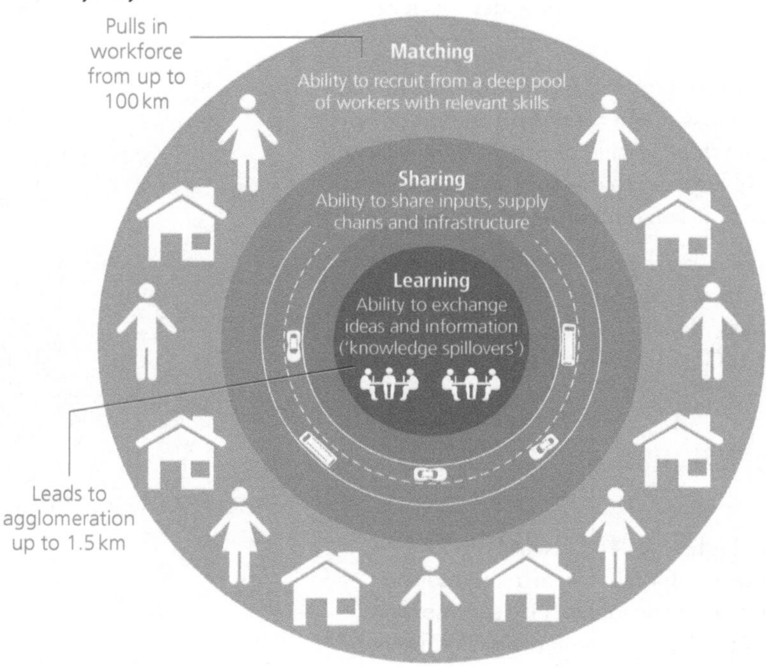

Figure 18 The benefits of agglomeration

- **Accessibility** allows for better shopping locations for lower-income groups because there is more choice and chances to compare. Out-of-town employment locations have fewer public transport connections and are therefore less accessible for the low paid.
- **Highly qualified labour pools** attract the skilled service companies, who in turn attract skilled workers. The share of graduates in Cambridge (36%), London (32%), Oxford (31%) and Reading (28%) in 2011 explains why these cities attract the highly skilled economic activities to their centres and nearby locations (**causality** and **difference**).

Is retailing declining in city/town centres and, if so, why?

UK town centres have undergone large changes in recent years. Retailing will only succeed if there are people who live and work in the city centre or if people are attracted to the city centre. The more retailing moves out of town, the fewer people will shop in the city centre. Changes in shopping habits, including competition from online services, business rates and the financial failure of some major retail chains have all impacted on the tertiary services found in the central parts of urban areas. In January 2019 the vacancy rate in town centres was almost 10% and retail footfall had fallen for the fourteenth consecutive month. Table 12 shows the changes in the businesses found in town centres, reflecting the changes in shopping habits. Many of the businesses that are closing are seen as town centre 'anchors', which would bring people into the area.

Table 12 Changes in business types in UK town centres 2013–17 (*The Grimsey Review 2*)

Top 10 business types opening	Change in number of stores 2013–17	Top 10 business types closing	Change in number of stores 2013–17
Tobacconists/e-cigarettes	+2,090	Banks and financial institutions	–2,405
Barbers	+2,066	Pubs and inns	–1,931
Beauty salons	+1,599	Women's clothes outlets	–1,588
Cafés and tearooms (independent)	+1,384	Newsagents	–1,357
Convenience stores	+1,185	Travel agents	–1,229
Hair and beauty salons	+986	Post offices	–1,087
Coffee shop (chains)	+981	Shoe shops	–861
Nail salons	+944	Chemists	–698
Restaurant and bars	+941	Fashion shops	–698
Takeaway food outlets	+902	Cheque cashing	–686

Knowledge check 12

Describe the trends shown by the information in Table 12.

Between 2008 and 2018 there were 32 main retail failures affecting 12,700 stores and over 175,000 employees. In 2018 Toys R Us, House of Fraser, Maplin, HMV and Poundland were all affected. Closed frontages, especially in larger stores, can alter the image of a high street. The replacement retail outlet might not return the street to its former image, causing the attractiveness of certain streets/centres to decline further. The loss of jobs may reduce the importance of a town centre as a place of work and tourism (**adaptation**, **attachment**, **difference** and **resilience**).

Many shops are increasingly becoming showrooms for browsing, making way for online and delivered-direct or 'click-and-collect' purchases. Table 13 lists some of the advantages and disadvantages of physical shopping. The reverse of these could easily apply to internet shopping.

Table 13 Physical shopping advantages and disadvantages

Advantages of physical shopping	Disadvantages of physical shopping
More pleasant and social	Opening hours are often limited
Can make unexpected, impulse purchases	Getting to and from shops, congestion and cost
Shopping as leisure — coffee bars	Parking and transport costs high
Individual service from approachable retailers	Less easy to compare prices in different shops
Direct possession and use	Pushy sales people
Feeling of reliability because of ability to make comparisons	More expensive due to the cost of space to store and display goods
Perceived as being more versatile because products can be compared	Lack of background information on products and services, which the internet can provide

E-commerce sales have steadily increased. By the start of 2019 over 18% of all total retail sales were completed online, accounting for £1 in every £5 spent. Table 14 shows the most popular items being purchased online by households.

Table 14 Most popular goods purchased online 2018

Goods purchased online	% of households who purchased
Clothes and sports goods	55
Household (furniture, toys etc.)	48
Holiday accommodation	42
Music/films	34
Books	31
Food/groceries	28
Electronics	27

Gentrification and associated social changes in central urban places experiencing reurbanisation

Table 15 shows that urban areas have continually undergone changes.

Table 15 The city centre timeline

Nineteenth century	Movement into cities during Industrial Revolution	URBANISATION
Early twentieth century	Business agglomerations and HQs concentrated in big city centres	AGGLOMERATION
1930s	Rapid suburban growth; car ownership rising	
1940s	Second World War bomb damage to centres — reconstruction necessary	
1950s–60s	Rising car ownership; abandonment of inner city for suburbs and estates	DECENTRALISATION
1960s–70s	Large-scale rebuilding, slum clearance, large commercial buildings; affluence rising	
1960s–80s	Decline of manufacturing and deterioration of industrial cities; office development restricted in London and spreads to other cities	
1970s–80s	Early gentrification in London (Notting Hill, Islington, Fulham); increasing car dependence; urban motorway schemes halted	
1980s	Fewer restrictions on commercial space outside of town centres; retail parks	
1980–2000	Expansion of tertiary economy and demand for office space	
1980s–90s	Urban regeneration — increase in retail, office and leisure space in CBD	REURBANISATION
1990s	Growing use of ICT and telephonic technologies	
1990–2000	Out-of-town development restricted; congestion as a result of more car travel; reurbanisation commences	
2008–2014	Recession, retail vacancies rise; internet shopping affects retailing	

Between 1971 and 1991 the number of people living in the centres of UK cities declined. However, since 1991 the population in city centres has risen, doubling between 2001 and 2011. The growth has continued, with the fastest growth being found in the bigger cities (Table 16).

Table 16 Fastest-growing city centre populations 2002–15 (Office for National Statistics)

City	Growth of city centre population (2002–2015) (%)
Liverpool	181
Birmingham	163
Leeds	150
Manchester	149
Bradford	146
Leicester	145
Sheffield	139
Newcastle	113

There are a number of factors resulting in the dominance of reurbanisation:

- **Building of flats and apartments:** more flats were built in Central Manchester between 2001 and 2011 than in the whole of London.
- **Permitted development rights:** these allow developers to convert offices to residential use in those areas where demand for residences is high.
- **Rapid population increase:** in Manchester and Sheffield the city centre populations grew by 198% and 111%, respectively, between 2001 and 2011.
- **Increase in student numbers at universities located in city centres:** 39% of university students in Manchester live in the city centre.
- **Increase in 25–34-year-old professionals in highly skilled jobs:** in larger cities they are more likely to be single graduates; in smaller cities they are family members often commuting into the city to work.
- **People live and work in the centre:** 39% of Central Manchester's population also works in the city centre.
- **Leisure and cultural facilities:** these are usually dependent on the young, educated population.
- **Cost of living:** in less successful twenty-first-century towns, such as Doncaster, Newport and Worthing, it is cheaper to live in the city centre.
- Young, single people and the elderly often have fewer possessions than families and can live in the smaller-than-average homes that are often found in the city centre. In 2016 7,800 micro-homes (less than 37 m^2) were built.

One consequence of economic and social change in central urban places is **gentrification**, where a poorer urban area improves in quality and becomes wealthier. The term frequently reflects the concentration of high- and ultra-high-net-worth individuals in the centre of major world cities (London, New York, Tokyo). London's ultra-high-wealth individuals had an average property portfolio of US$28 million in 2014. The reasons why many of these people invest in Central London are as follows:

- Property investment gives high returns due to inflation and exchange rate changes that favour the investor.
- Demand for accommodation — the UK is a safe place for investment.
- They wish to have a home for their globalised lifestyle.

Gentrification is associated with the service economy and sees highly educated professional, creative, technical and managerial workers replace industrial workers. Property developers with financial backing also convert older industrial buildings into apartments in these areas (**causality**, **adaptation**, **risk** and **resilience**).

- The older industrial buildings are often converted into luxury apartments rather than social housing and so attract high-income, young professionals.
- High-income earners can afford to renovate the properties in older areas.
- Wapping, London was the first area in the city in which waterside warehouses were converted into apartments. This trend has spread to other areas of London where former industrial buildings have been abandoned.
- More recently brownfield sites, such as Battersea Power Station and areas of poor-quality buildings around it, are being cleared and developed into luxury high-rise apartments.
- In areas such as Shoreditch and Hoxton — former nineteenth-century working-class residential districts — houses are being converted to house the highly paid city employees. The 1902 Brune Street Soup Kitchen has been converted into luxury flats. Edward England Wharf in Cardiff is a former potato factory that is being converted into apartments. Even in rural places, such as Arundel, the nineteenth-century Poor House has been converted to flats.
- As areas become more desirable, this can result in increasing house prices, which only the wealthy can afford.

A 2015 survey highlighting the reasons why people lived in the centres of Manchester and Brighton is summarised in Table 17 (**identity**, **meaning** and **representation**).

Table 17 Reasons for people living in the centres of Brighton and Manchester (percentage of those surveyed who gave a reason)

Brighton	Reason	Manchester
30%	Close to restaurants, leisure — the centre as a central entertainment district	60%
15%	Close to workplace	40%
28%	Public transport	32%
15%	Type of housing available	5%
23%	Cost of housing	26%
4%	I grew up here	1%

These people are the gentrifiers who work long hours and need to live close to their work. They also want the cultural and entertainment opportunities that a city centre provides.

The complexity of the changing service economy

All urban areas grow and decline as a result of changing social and economic forces over time. The service economy has resulted in population growth and changing lifestyles in all urban areas, but mainly in the southeast of the UK. Between 1971 and 2009 population growth in towns and cities was rapid (Milton Keynes 253%, Telford 67%, Crawley 29%, Cambridge 47%, Reading 34%).

On the other hand, during the same period, deindustrialising urban areas were losing people and economic activity (Liverpool –19%, Tyneside –10%, Stoke-on-Trent –4%, Burnley –3%). Demand for space and for new buildings in these places is low and, in the core, dwellings may be vacant/derelict. In declining cities there are attempts to improve the quality of the built environment and people's neighbourhoods through **remediation** (clearing derelict sites), which aims to bring the land back into use as offices, business parks, leisure, tourism and housing. Many such top-down schemes have limited success and have not created the anticipated number of jobs. They have come to depend on local government financing (e.g. Barnsley Gateway) because there is little demand for commercial property.

Furthermore, because the emphasis is mainly on commercial regeneration, the effect on people has been largely negative. 'The starting point for any serious urban policy is to recognise that the objective should be to enrich and empower the lives of people, no matter where they live' (Ed Glaeser 2008). For some cities, 'smart decline' and creating more green space would be better options (**difference** and **mitigation**).

Figure 19 shows urban employment in two contrasting places.

Knowledge check 13

Comment on the method of presentation used in Figure 19. How else could the data be shown?

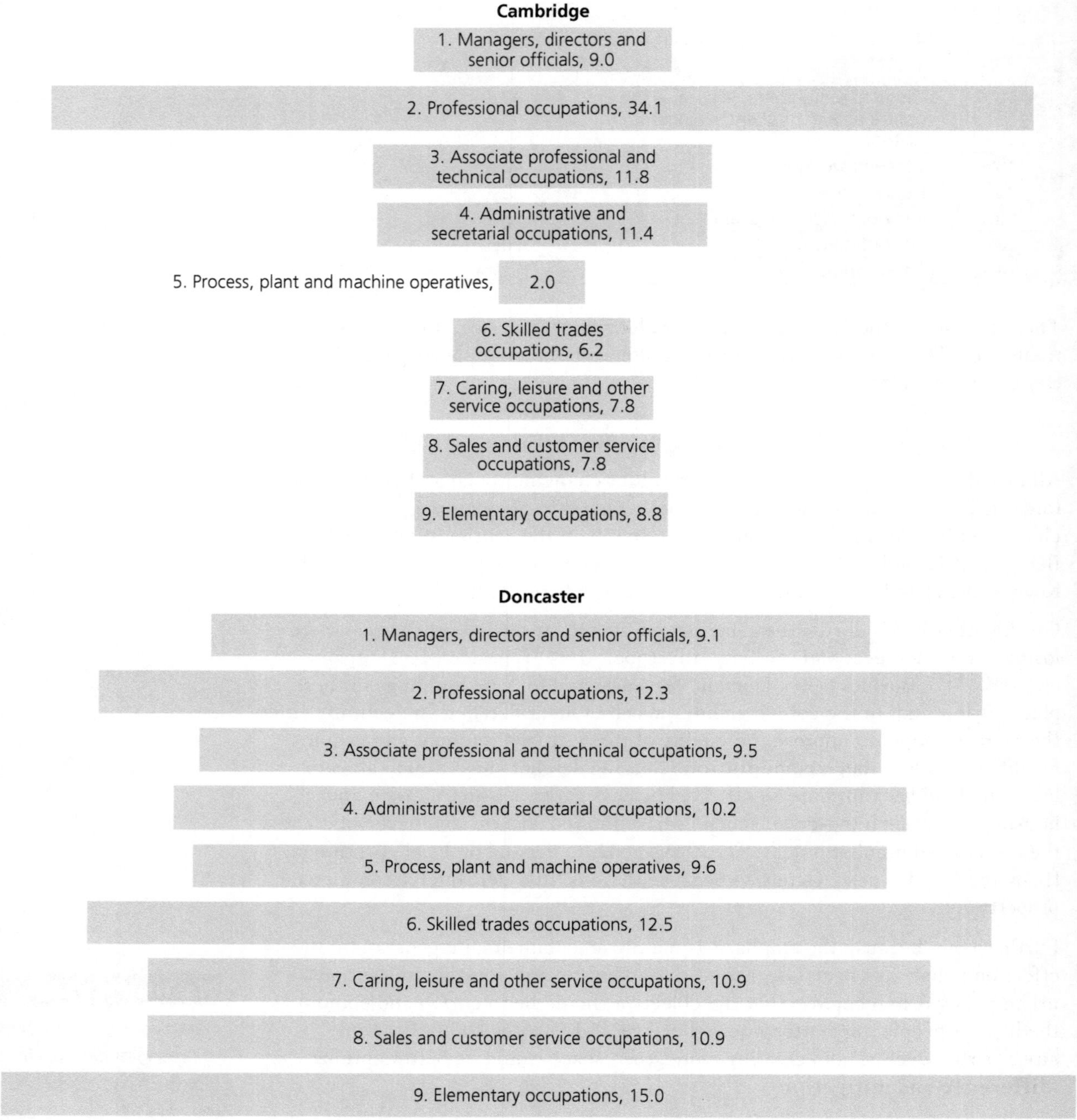

Cambridge

1. Managers, directors and senior officials, 9.0
2. Professional occupations, 34.1
3. Associate professional and technical occupations, 11.8
4. Administrative and secretarial occupations, 11.4
5. Process, plant and machine operatives, 2.0
6. Skilled trades occupations, 6.2
7. Caring, leisure and other service occupations, 7.8
8. Sales and customer service occupations, 7.8
9. Elementary occupations, 8.8

Doncaster

1. Managers, directors and senior officials, 9.1
2. Professional occupations, 12.3
3. Associate professional and technical occupations, 9.5
4. Administrative and secretarial occupations, 10.2
5. Process, plant and machine operatives, 9.6
6. Skilled trades occupations, 12.5
7. Caring, leisure and other service occupations, 10.9
8. Sales and customer service occupations, 10.9
9. Elementary occupations, 15.0

Figure 19 Employment by occupation groups in Cambridge and Doncaster 2011

Functions: why, where and examples

The growth of the service economy over the past 50 years has seen a wider variety of locations selected and occupied than ever before.

Retailing (out of town)

- Rise of superstores, e.g. Sainsbury's, Tesco, and new supermarkets, e.g. Aldi and Lidl.
- DIY boom: either purpose-built or in converted industrial units, e.g. Homebase, IKEA.
- Electrical and electronic goods, e.g. Dixon Group.
- Furniture and homeware, e.g. Furniture Village, DFS, John Lewis.
- Clothing, e.g. Next.
- Combined sites to increase customer flow, often adjacent to major junctions, e.g. M&S and Sainsbury's, Hedge End Southampton (**interdependence**).
- Outlet stores, e.g. Bicester Village, Bridgend Designer Outlet, Cheshire Oaks Designer Outlet.
- Retail-attached service stations and service areas on motorways, e.g. M&S, Waitrose.
- Industrial estates — many date from earlier in the twentieth century but the newer industries are service orientated, replacing small former industrial units, e.g. double glazing, car wash.

Exam tip

Always try to support your answers with located examples, such as some of those given here.

Retailing (city centre)

- City centre regeneration, e.g. WestQuay, Southampton; St David's Dewy Sant, Cardiff; Westfield London, Hammersmith.
- Creating a new city centre, e.g. Westfield Valley Fair Mall in San José, California has a new open-air shopping and leisure (hotels and restaurants) street, Santana Row, leading from it, which creates a more traditional image of a city centre (**interdependence**).
- Regeneration of redundant industrial space, e.g. Gunwharf Quays, Portsmouth.
- New convenience stores, renovated pubs, and corner shops within the inner suburbs, e.g. SPAR, Tesco Express.
- Car showrooms on routes into city/urban area, e.g. BMW, Audi.
- Specialist shops in small towns, e.g. fashion, cheese, booksellers, antiques.
- Charity shops, loan shops, betting shops, usually in smaller centres and streets leading to a centre.
- Mainly internet retailers, which are not tied to a location. This may affect the size of stores for some of the retailers because they do not have to stock the products on site. Instead they showcase/demonstrate the products, which are then bought online.

Offices

- Suburban clusters, e.g. Canary Wharf, London.
- Barn conversions to form small office parks in rural areas (**resilience**).
- Office parks, e.g. on the reclaimed Port Solent, Portsmouth (began when IBM's offices were decentralised from London) (**space**).
- University-developed science parks, e.g. Cambridge, Southampton, Stanford Research Park, San José, California (**interdependence**).

Leisure

- Leisure quarter in UK city centres often in converted buildings.
- Restaurants taking over from former uses, e.g. Jamie's, Cambridge and Zizi, Cardiff in former banks (**adaptation**, **resilience**),
- Multiplex cinemas both on CBD fringe and out of town, often in combination with retail developments (**interdependence**).
- Multi-purpose leisure centres, e.g. Fleming Park Eastleigh.
- Visitor centres, museums and galleries in converted or regenerated buildings, e.g. San Francisco Museum of Modern Art (SFMOMA) (**adaptation**).
- Hotels, near entry points to city and on motorways, e.g. Premier Inn, Ibis.
- Hotels as part of regeneration, e.g. St David's Hotel, Cardiff Bay; Ghirardelli Hotel, San Francisco, in a former chocolate factory (**adaptation**).
- Conference centres, e.g. BIC and International Convention Centre (ICC), Birmingham; Santa Clara Convention Centre, California (**globalisation**).
- Exhibition centres out of town, e.g. National Exhibition Centre (NEC) Birmingham; Excel, London; San Matteo Event Centre, California (**globalisation** of products and activities).
- Stadia as concert venues, e.g. Wembley, London; Principality Stadium (formerly Millennium Stadium), Cardiff; Levi Stadium, Santa Clara, California (**globalisation** of sports and stadium concerts).

All these developments do have *risks*. Can you relate the key concepts to all of the bullets above? (There are several examples already included as a guide.)

Exam tip

Always try to have a bank of examples from beyond the UK to demonstrate your breadth of knowledge. The non-UK examples above are all from the Silicon Valley area of California — a contrasting place.

How has central city retailing changed? Not all of the reasons apply everywhere, but examine how they apply to your home town (**difference**, **adaptation** and **sustainability**):

- More specialisation resulting from new products, such as mobile phones.
- New retailers replacing those that left, e.g. Primark in old Woolworths stores.
- New large-scale retail developments with anchor stores such as John Lewis, St David's, Cardiff and Grand Central, Birmingham.
- Lower turnover of shops on side streets due to their lack of visibility or footfall.
- Greater affluence, with designer brands that sometimes cluster, e.g. The Hayes, Cardiff.
- New technologies, e.g. 'click and collect'.
- Rising car ownership, but costs of parking and congestion can deter visitors.
- In the less frequented and peripheral streets, the rise of poorer-quality shops such as betting shops/casinos, charity shops and loan/pawn shops. These can take over the heart of some less prosperous towns.

Impacts of tertiarisation on employment

The rise of the service-based economy has been accompanied by the growth of professional, managerial, technical and creative employment of highly educated and highly paid persons.

Figure 20 shows how employment in two cities polarised between 2001 and 2011 in the light of the trends shown in Figure 21. In Stoke-on-Trent intermediate jobs declined by 8% (12.6% drop in manufacturing jobs), whereas in Peterborough the

Knowledge check 14

What potential occupational trends does Figure 20 show?

decline was **mitigated** by a rise in industrial operatives. In both cities the low-wage sectors, especially the care sector, grew far more than the high-wage sectors.

Figure 21 shows that the balance of occupations is changing as the tertiary sector expands.

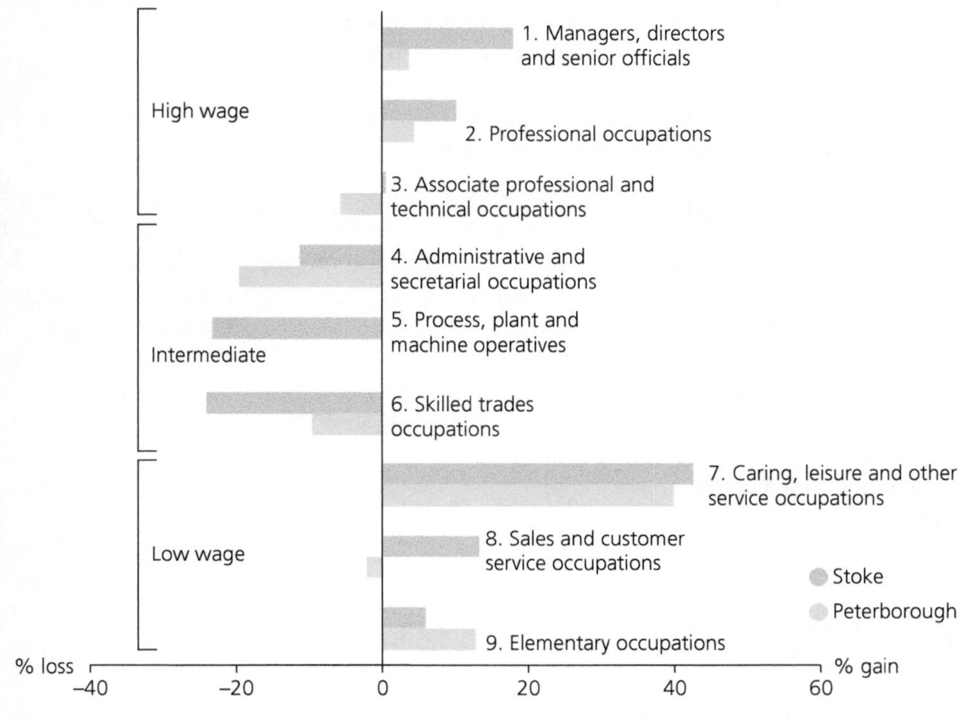

Figure 20 Changing occupational structure in Peterborough and Stoke-on-Trent 2011

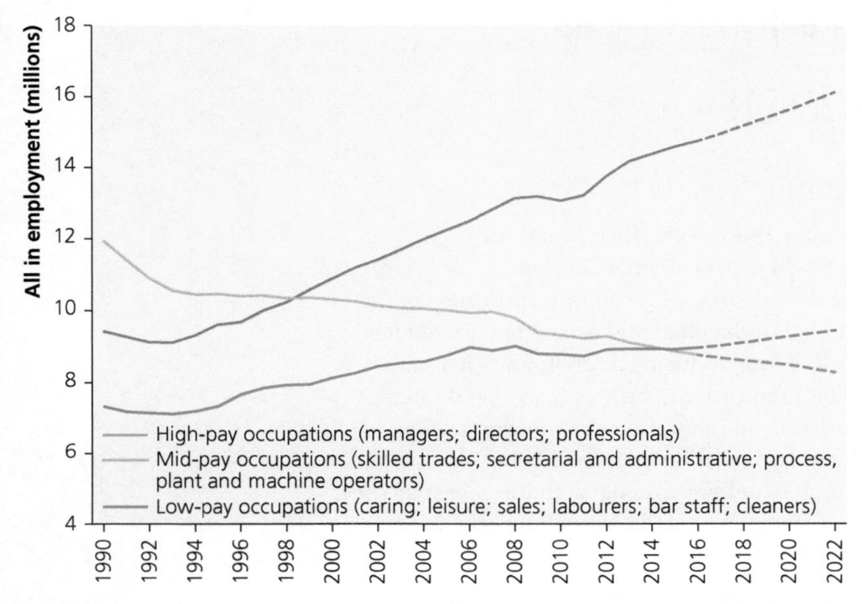

Figure 21 Changes in high-, middle- and low-wage occupations in UK cities 1990–2016, and prediction to 2022

Tertiarisation has also resulted in the growth of low-paid occupations, such as restaurant/bar staff, cleaners and the leisure industry. Combined with the impact of an ageing population, the care sector's employment has greatly increased. Between 1990 and 2012 the care sector's employment had tripled. In 1990 3.3% worked in the sector; this rose to 6.9% by 2012. Since 2009 the care sector workforce in England has increased by 21% to 1.6 million. More than 55% of these jobs are in cities, and they are low paid and subject to the effects of recession and austerity. In 2014 17% of care-sector workers were in 'work poverty' (8% nationally for all workers). Many jobs are part-time, involve flexible working and are taken up by women. Over 25% are on zero-hours contracts. The 2020 coronavirus pandemic has drawn attention to this sector, which may have future implications.

Summary

- Service-sector growth has characterised places in advanced economies for over a century.
- Urban centres have been the traditional locations for service-sector growth.
- City centres have changed in form and functions to include, offices, retailing and leisure.
- In the late twentieth century service sector employment spread to other locations within and beyond urban areas.
- Retailing as a city-centre service has been challenged by out-of-town and internet shopping.
- Reurbanisation and gentrification are altering the identity of central areas.
- Some central area developments are on former industrial/port areas.

The twenty-first-century knowledge economy (quaternary) and its social and economic impacts

What is the quaternary economy?

The quaternary economy forms the major activity of the fifth Kondratiev wave (Figure 5, page 20) and is an economy based on the creation, evaluation and trading of ideas and information. It is characterised by the rise of communication and computer/information technologies, and closely allied to the changing nature of communication and transportation systems, which together are producing a set of new townscapes and changed places. Despite modern communications, such as high-speed internet, clustering is occurring because productivity and innovation are concentrated in cities. Innovation needs backers, such as city-based financiers and entrepreneurs. Cities have a high density of creative, digital and professional activities because they have the threshold to support the concentration of a skilled workforce, the existing broadband infrastructure and international trade links.

Intra-industry spillovers are where the proximity of similar firms enables knowledge and ideas to travel among specialist companies, furthering the development of new activities. In 2015, among the smaller players (small- and medium-sized enterprises, or SMEs), 65% of the creative industries and 60% of the digital companies were located in cities.

The digital economy is expanding 2.6 times faster than the rest of the UK economy and is worth £184 billion to the economy, creating 31% of the UK's GDP. One fifth of the technology workforce is located in London.

What follows is a list of terms related to the quaternary economy:

- The **knowledge economy** is founded upon the ability of people to innovate, which is at the heart of technological change.
- **Knowledge-intensive business services** (KIBS) are still highly localised despite the ability of IT systems to disseminate knowledge to the world. Knowledge tends to be produced in specific places, such as Silicon Valley or Cambridge, and used and improved upon in the same location. 50% of KIBS jobs in city centres are taken by graduates.
- The **digital economy** employed over 11% of the UK workforce (2.1 million people) in 2018. It includes ICT (software development, broadband networks, hardware, software, sales and marketing) and digital content (digital media, publishing, design, music and advertising).
- The **creative industries** include product design and software development, broadcasting, advertising, libraries and museums.
- **Biotechnology** and other scientific developments, such as medical research, together with legal, accounting and management consultancy, are sometimes included.
- **Unicorns** are companies that have recently started up and are already worth over US$1billion, for example Uber and Airbnb, both of which originated in Silicon Valley. Of the 13 UK unicorns, nine are based in London, for example Deliveroo, which is located in Islington.
- One new job in the digital knowledge economy leads to five jobs elsewhere in the economy, a feature known as the **multiplier effect**.

One way of measuring innovation and the strength of the knowledge economy is by the number of patents granted (Table 18). It is possible to see the link between the number of patents and the proportion of the population with higher-level qualifications. At the national level, China, South Korea, Japan, Germany and the USA dominate.

Knowledge check 15

Describe the differences in the distribution of UK cities/towns with the highest number of patents per 100,000 and the cities with the lowest.

Table 18 Patents granted per 100,000 population and proportion of population with high-level qualifications 2018

Cities/towns with highest number of patents granted			Cities/towns with lowest number of patents granted		
City/town	Patents granted per 100,000	% of population with high-level qualifications	City/town	Patents granted per 100,000	% of population with high-level qualifications
Cambridge	269.8	58.1	Barnsley	8.3	31.8
Coventry	113.3	34.7	Hull	7.3	26.7
Oxford	93.5	63.0	Glasgow	7.1	47.4
Derby	81.1	31.9	Sunderland	6.9	27.4
Aberdeen	55.5	51.7	Burnley	6.6	27.8
Crawley	49.1	33.2	Doncaster	6.6	23.6
Aldershot	46.6	49.6	Southend	6.4	26.4
Edinburgh	33.7	57.8	Mansfield	5.8	17.8
Bristol	30.6	49.4	Luton	4.1	33.9
Slough	29.8	39.4	Wigan	3.1	26.8

Locational factors encouraging cluster growth

In the UK a number of factors have led to cluster growth (Table 19).

Table 19 Factors encouraging cluster growth

Factor	Description
Government support	The government has announced an increase of 50% in R&D by 2027 National and local government support was given to East London Tech City
Role of universities and research establishments	The government allows the universities to establish local growth plans and University Enterprise Zones, with business spaces for new high-tech companies starting up Presence of a highly educated workforce needed by companies and for start-ups Encourage investment in R&D
Planning regulations	EZs and LEPs (page 36) encourage new industries
Infrastructure	Industry locates in areas with good connectivity to other parts of the country and globally

Many countries have areas of knowledge economy concentrations, including: Silicon Alley, Lower Manhattan, New York (stretching from the Flatiron District to Soho); Cap Digital and Silicon Sentier in the former textile area of Paris; Silicon Wadi in Israel; and MSC Malaysia, formerly the Multimedia Super Corridor, near Kuala Lumpur. Perhaps the best-known among them are Silicon Valley, California, and Silicon Roundabout, in the Old Street area of London. At a smaller scale, the Menai Hub is the promotional name given to the cultural and knowledge-based activities in Bangor, Caernarfon and Llangefni in North Wales.

Silicon Valley, California

Silicon Valley, once called 'the valley of heart's delight' (Figure 22), extends the length of the Santa Clara Valley between San Francisco and San José in California. Until the 1940s it was an area dominated by fruit orchards. Today, it is a set of low-density cities and towns linked by freeways and the Caltrain.

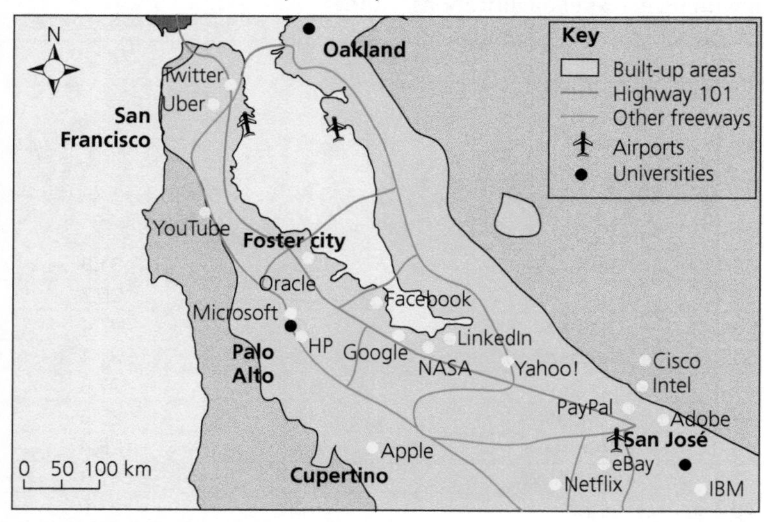

Figure 22 Silicon Valley

Knowledge check 16

Why might a large number of patents granted reflect an increasing reliance on the knowledge economy?

Exam tip

Clusters can lead to improvements in global connectivity, which may attract further investment, resulting in the development of a **global hub**. This topic forms part of the Global governance unit of Eduqas A-level Component 2.

The factors (**causality**) that gave rise to the growth of the knowledge-based economy in the Valley are as follows:

- **Triggers:** the decision by Hewlett-Packard to develop oscillators in their garage in Palo Alto is one starting point. Before that, the area had attracted radio enthusiasts, who began to create an electronics community. The peninsula offered lifestyle benefits to the early innovators; the ocean and the mountains were within easy reach and the Mediterranean climate enabled the development of an outdoor lifestyle that appealed to the young. Until the 1960s property was cheap and surrounded by orchards, grasslands and forested hills.
- **The role of Stanford University:** its research professors and students have been a constant source of ideas and innovations that have spawned many of the companies that are now found in the region. Other universities in California have also played their part in providing researchers and developers, such as San José State, Berkeley and Caltech (**attachment**, **causality**).
- **Inspiration of individuals:** Terman, the Dean of Engineering, set up the Stanford Industrial Park in the 1940s, which has attracted over 150 firms. He nudged Hewlett-Packard towards making oscillators. Shockley set up a laboratory in a former fruit-packing plant in Mountain View in 1956 to further develop transistors. Employees of Shockley became dissatisfied and broke away to form Fairchild Semiconductors, which in 1959 produced what would evolve into the silicon chip. One employee set up Intel in Santa Clara, which has become the world's leading chip developer and manufacturer. Other notable individuals are Brin and Page, former students at Stanford, who worked with Filo, the founder of Yahoo, and in 1998 founded Google, now a 13,000-employee campus in Mountain View.
- **Family trees of companies:** Fairchild became the parent from which offspring hardware companies (Apple, Cisco Systems, Sun Microsystems, Silicon Graphics) emerged. Google, Yahoo, eBay and Netscape piggybacked on the cluster of hardware companies.
- **Military needs:** during the Second World War and 1950s military needs resulted in the development of the Ames Research Center at Moffett Field, which evolved, partially as a result of large government contracts, into a NASA base (**interdependence**).
- **Young entrepreneurs:** these work for major IT companies where they subsequently develop their own ideas, which they then endeavour to take to the market. They require finance from venture capitalists. In the 1960s the venture capitalist industry developed close to Stanford University in Palo Alto and subsequently elsewhere in the region. These firms clustered close to their clients. Thus finance, just like talent, agglomerated in the Valley (**adaptation**, **interdependence**, **risk**).
- **Talent refusing to leave the area — inertia:** other IT giants arrived because the talent they needed to recruit refused to leave the Bay Area, for example to go to Microsoft's HQ in Redmond, Washington State. Microsoft recruited a guru named Gray who insisted that the Bay Area Research Centre (BARC) be opened in San Francisco in 1995, which expanded more recently into Mountain View (**attachment**, **identity**).
- **Transport provision:** both regionally and internationally, transport provision has assisted development. Freeways such as the 101 regularly slow because of the sheer volume of commuters, but the Caltrain line running through the Valley has enabled

a myriad commuting patterns between locations. In addition, the presence of major airports, such as San Francisco International, has made it easier to connect with the global economy. There are now more large airports (Oakland International, Mineta San José International), and also small airports such as San Carlos, where the CEO of Oracle has his private jet located a mere 5 minutes away from the office (**interdependence**).

- **Ability of firms to recruit talent from around the world:** this applies especially to Asian software engineers (page XX). The range of highly qualified employees (58,000 new jobs were created in 2014 alone) adds to the positive vibes and the growth of success, and increasingly diverse amenities, restaurants and social provision available in the area. This in turn attracts more talent to the agglomeration. Globalisation becomes easier when much of your workforce originates from countries around the world (**globalisation**).
- **Takeovers:** these are another means by which firms get larger, for example Google acquired YouTube, which is located in San Bruno.
- **Other organisations moving to the Bay Area to capture talent:** because Silicon Valley and San Francisco Bay are the places to find talent and for new talent to be nurtured, other organisations are moving to the Bay Area. Tesla, the electric car maker, has its main plant in Fremont, and Airbus announced in 2015 that it is to develop a research facility in the area, rather than in Europe, in order to make use of the expertise. In both of these cases the attraction is the availability of software engineers who can develop improved systems for the modern car or airplane. Biotechnology firms have also used graduate concentration to help them expand in the area. The DOE Joint Genome Institute at Walnut Creek was an early locator, and now firms such as Gilead Sciences have expanded rapidly in Foster City. Major financial organisations, such as Visa in Foster City, have arrived to latch onto the talent pool. Educational innovators are also attracted to the area; in 2016 a French organisation announced that it was setting up a free university in Fremont to train IT workers (**causality**, **feedback**, **globalisation**). Table 20 summarises the costs and benefits of Silicon Valley as a **place**.

Table 20 Costs and benefits of the Silicon Valley agglomeration

Costs	Benefits/reactions
Large-scale migration from within the USA and from overseas leads to the following: ■ Demand for housing, schools and facilities for young families ■ High price of property due to demand exceeding supply, therefore employees are forced to live further away, where prices are lower ■ Transport system overloaded, not only in rush hour, due to the 24/7 working pattern ■ High levels of pollution ■ Water supply issues, especially with summer droughts and extended drought periods (2012–2015) ■ Older residents feel that the environment and places that they grew up in are being destroyed by progress ■ Space to build disappearing rapidly, forcing movement onto protected land ■ Difficulty for public service workers, such as teachers, to be recruited because of the high price of housing ■ Congestion	A cosmopolitan society, whose talents can be engaged in a variety of projects Excellent higher education that attracts quality students and researchers Developers building as fast as possible, with guaranteed profits; landlords also able to raise rents above inflation Companies providing own transport that enables work patterns to be more flexible Has led to high levels of hybrid car ownership and the rise of the electric car industry (Tesla) Environmental awareness is widespread The benefits of agglomeration and clustering for transfer of ideas and career progression Value placed upon spin-offs and new entrepreneurial ventures High levels of highly qualified posts for women The potential to poach quality staff from competitors — knowledge spillover

Tech City (Silicon Roundabout), Shoreditch, London

Between 2010 and 2016 the most significant locational growth of technical businesses (92% increase in digital firms in 2010–2013) has been in Inner London, where 300,000 now work in digital employment. In 2010, Prime Minister David Cameron gave the greatest concentration area its name — Tech City — to help turn a start-up area into a digital hub by attracting investment and talented entrepreneurs from all over the world. Also known as Silicon Roundabout, it is an inner urban cluster (**scale**) located within this global city. It is on the fringe of the City CBD and at the eastern end of a high-tech corridor that stretches to the West End (Figure 23). It has a mix of ICT and digital content sectors, with the heart of the area housing almost 6,000 tech companies since 2016. The cluster is centred on Old Street roundabout and extends into Hoxton and Haggerston, the City of London, Farringdon and Bethnal Green. The concentration is small compared with other sectors and other wards in London. However, it is an area with a high number of start-ups, especially in the arts and cultural services (**interdependence**).

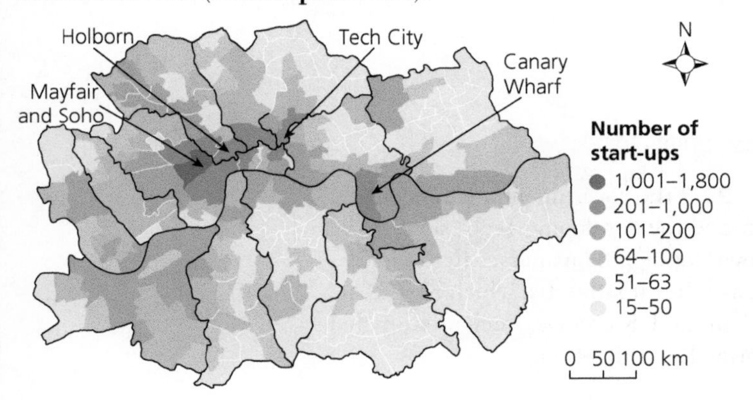

Figure 23 The geography of start-ups in London (2013)

The factors (**causality**) leading to the growth of Tech City are as follows:

- The amenities of the area, which cater for the new workforce — for example, café culture and the resultant vibes (**attachment**, **identity**, **meaning**).
- Similar, complementary firms that have clustered in the area; they make use of incubator space in converted Victorian warehouses (**adaptation** and **interdependence**).
- Branding and messaging have made both entrepreneurs and financiers aware of the new activities (**causality**).
- The renting of floor space in the area is far cheaper than in the CBD (**causality**).
- Proximity to central London and the City, which is the marketplace for the innovations and the source of finance (**interdependence**).
- Connectivity to the rest of London and the UK.
- Higher broadband speeds than in other parts of central London.

One of the effects of Tech City's popularity is an increase in housing rental costs — doubling between 2011 and 2013 — in the gentrified area. Increasing demand meant that, by 2018, a two-bedroom apartment in the area could cost £2,400 a month to rent. This compares with the average monthly rent in England for a two-

Exam tip

An internet search of images past and present of the Shoreditch area will show the area's new identity and how it has changed over the years. This will help you identify the impacts of the knowledge economy and gentrification.

bedroom house of £820, although the average for London is £1,700. Therefore, some entrepreneurs will be forced to look for new, cheaper locations in which to base their start-ups. This fact, plus a lack of availability of new properties, has caused the number of start-ups to decrease in recent years.

Transport links will influence the location of newer clusters, such as near King's Cross, where Google, Central St Martins School of Art and Macmillan Group have located on a former goods yard, thus drawing in other new technology ventures (page 77).

Figure 24 illustrates where clusters grow, often in areas that are already the focus of that particular activity. Activity in finance, culture and the arts is very concentrated, whereas professional services and ICT are more dispersed.

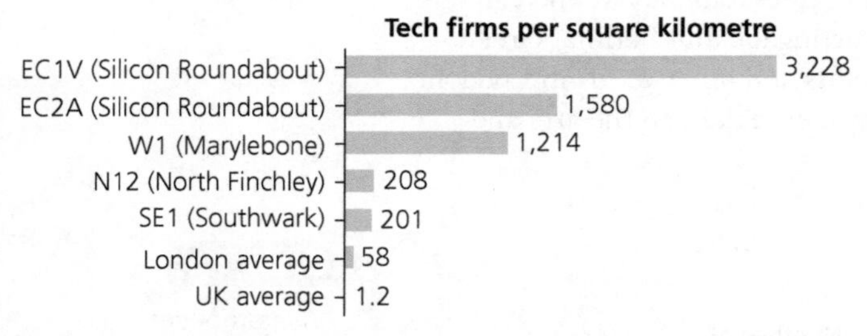

Figure 24 The main areas for start-ups in London 2008–2011

Government policy can influence clustering. In 2017 the government announced the launch of Tech Nation, consolidating Tech City and the developing Tech North, as a national network aiming to accelerate the growth of digital businesses in the UK. It has identified the growth of 16 areas known as **silicon suburbs**, which have a higher proportion of digital tech employment than the UK average. Such areas include Newbury, Reading, Basingstoke, Burnley, Huntingdon and Telford.

Education and its impact on cities

If you live in a university town you are probably very aware of the impact of the student population (including the employees of education establishments) on both the housing market and the provision of services.

Birmingham has over 65,000 students studying at five universities while Manchester has over 99,000 students at four universities. Other cities also have a large student population. While many students live in halls of residence, some of which are privately rented, a large proportion are housed in the private rented sector. We have already seen how students play a key part in the gentrification of city centres in their new halls of residence (page 43). After graduation, many students may stay in the area, helping the process of gentrification, and become involved in new start-up businesses in the knowledge economy.

Impacts of quaternary industry clusters on people and places

Business and science parks have developed around many towns and cities. Cambridge developed the first science park in 1973. Thirteen such sites now function in up to a 10-mile radius around the city. The university and highly skilled graduates combined with the conserved historic core have attracted KIBS jobs to the city and pushed up living costs. Even within the city, a new knowledge industry space, called CB1, is being developed around the railway station, which has already attracted Microsoft and will house the international examination board Cambridge Assessment. Some cities have over-supplied these locations, for example Newcastle-upon-Tyne, often using their EZ status (page 36), and several sites in the city remain under-occupied.

The impact of quaternary activity on people, localities and places is summarised in Figures 25 and 26. **Digital exclusion** refers to a lack of skills associated with computers, together with poor access to broadband. Rural areas (e.g. Northumberland, Anglesey) are most likely to exclude people due to the lack of access to broadband, because there is not the threshold population to support the investment in cabling in these regions. In 2019, 9.3% of adults in Wales had never used the internet. The reasons for low use of digital technologies are varied and include a more elderly population, out-migration of young adults to universities, no training for the older generation and a lack of jobs in the area.

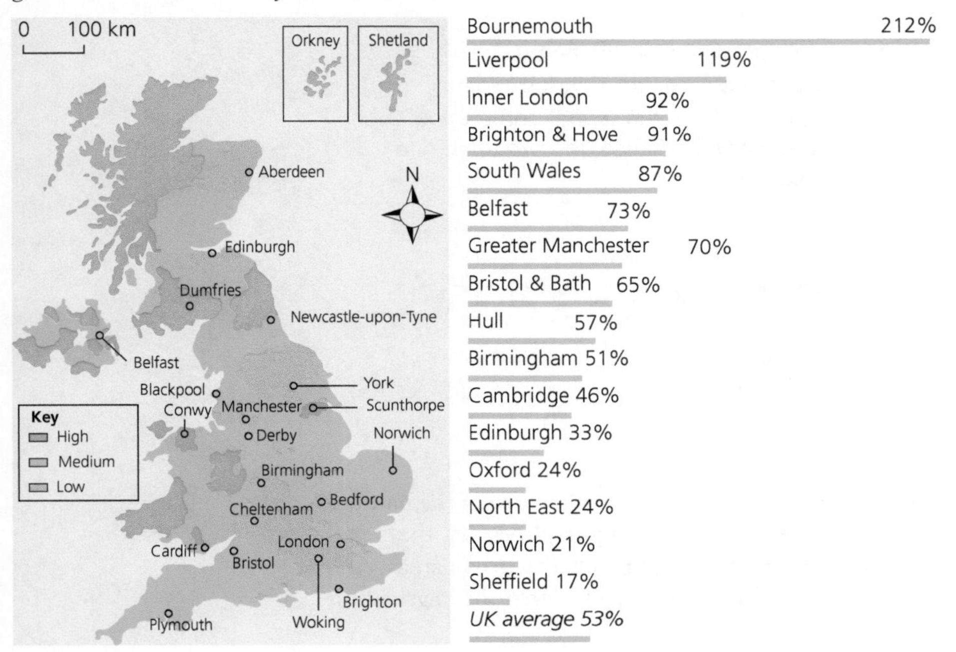

Figure 25 Digital exclusion in the UK

Figure 26 Digital company growth 2010–2013

Urban areas (e.g. Woking, Guildford, Southampton, Aberdeen, Warwick) are less excluded from broadband, resulting in low levels of exclusion overall. However, there are some urban areas, such as north Lincolnshire around Scunthorpe and Rhondda Cynon Taf, where having access to broadband, low educational attainment and relative poverty can lead to social exclusion (**inequality**).

South Wales, especially Cardiff, has seen a growing number employed in the tech sector (28,000 in 2015). Proximity to the Welsh government in Cardiff Bay and government support via the Business Wales Digital Development Fund have helped. The staff and space are cheaper than elsewhere in the UK, which has aided the establishment of the Innovation Centre for Enterprise (ICE) in Caerphilly, one of the most deprived towns in Wales (**mitigation**).

Knowledge check 17

Use Figure 26 to describe and offer explanations for digital company growth. Are the data displayed in the best format?

Clustering can attract a higher-educated, digitally proficient workforce earning wages 36% higher than the UK average. This can change the social character of an area, creating changes in services as well as creating a demand for homes, resulting in increasing prices. Non-quaternary workers earning lower wages may be socially excluded from such areas.

Summary

- Quaternary activities are of increasing importance in both urban and rural places.
- The factors determining the location of quaternary activity include some that were important for secondary industry, such as clustering, but also factors that are related to the importance of new technologies and entrepreneurship in the twenty-first century.
- Educational hubs and research centres are important attractions for digital and biotech companies.
- Quaternary, knowledge-based activities are being driven by large multinational companies and, at the same time, small independent companies/individuals that have spun off from the global companies.
- Some people are excluded from these new activities and are unable to benefit from the use of new technologies.

The rebranding process and players in rural places

Rural places

You should distinguish between **rurality**, the degree to which an area of the natural, non-urban world depends on agriculture/food/forestry, and **peripherality**, the distance either in time or space from the opportunities provided by urban areas. A total of 80% of employment in rural areas in England is in jobs other than agriculture/fishing (7%) and tourism (12%).

Since 2004 rural areas have been defined as areas in which no settlement is greater than 10,000 people. Rural areas are divided into sparse (low density of dwellings) and less sparse (higher density of dwellings — see Figure 27). Those areas classified as sparse are not necessarily what a geographer would class as remote rural, although some are remote.

A different classification (Figure 28) recognises that urban influence and access extend to large areas around cities, either as 'suburbs' or 'hinterland', all of which extend well into the countryside. Very little rural land is left.

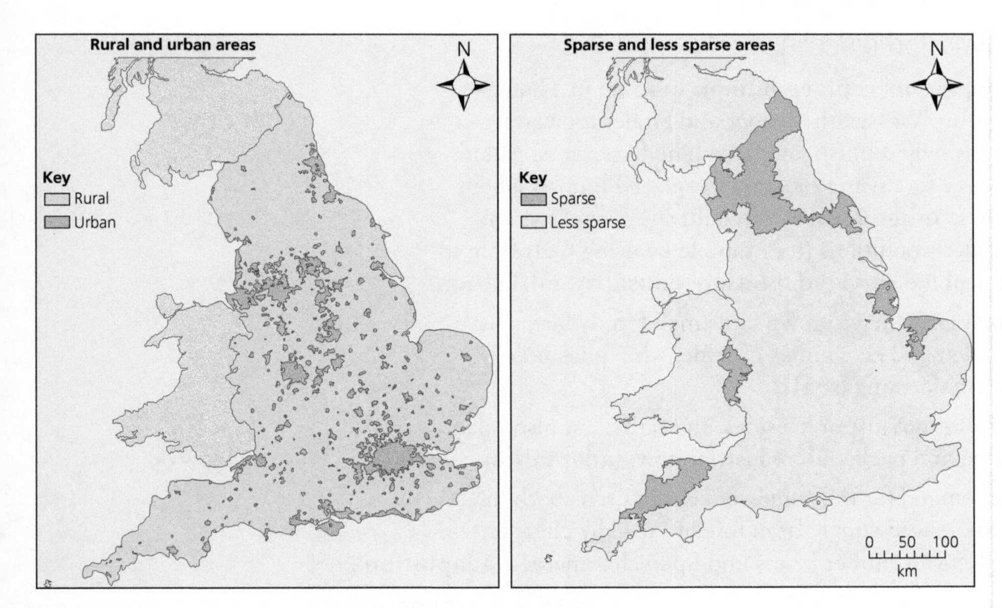

Figure 27 Rural/urban classifications in England

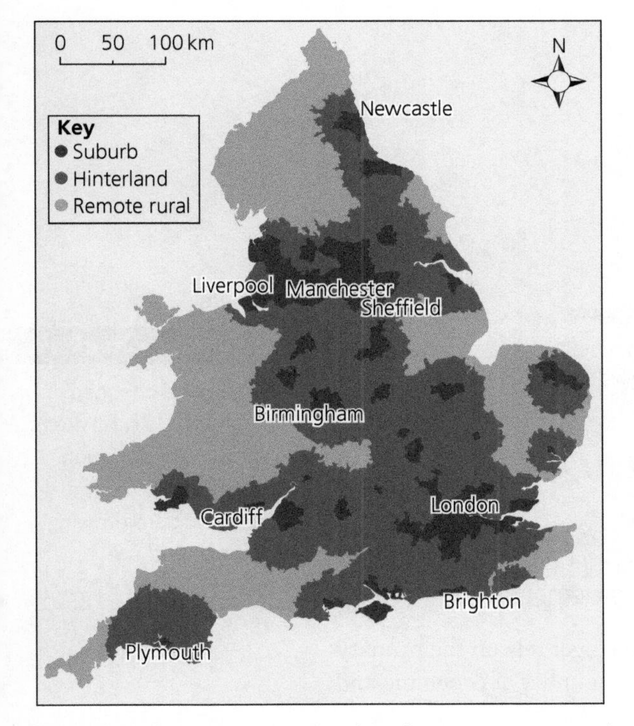

Figure 28 Suburbs, hinterland and more remote rural areas (2015)

A 2.6% increase since 2011 meant that in 2017 17% of the population (9.4 million people) lived in rural areas. Of these people, only 7.7% were employed in agriculture, forestry and fishing. Figure 4b (page 20) shows how the percentage of the workforce in primary industry has steadily decreased to just 1%.

Exam tip

Geographers use the term rural–urban fringe, but this encompasses several of the government definitions. Be aware of these variations in terms, which can affect decisions made about places.

Why has the primary workforce declined?

1 **Technological developments replace human labour:** in Thomas Hardy's nineteenth-century Wessex the reaper and binder took away jobs; last century, increasingly sophisticated combine harvesters, picking technology and machinery for raising root crops replaced human labour. There has also been a rise in agriunits, which cultivate crops (such as tomatoes) under controlled conditions (for example by using CO_2 from the ICI Teesside chemical industry) (**adaptation**, **causality**, **mitigation**).

2 **Rising scale of farms:** farms have grown and small family farms have merged. Supermarkets demand economies of scale, which has been especially evident in dairy farming (**scale**).

3 **Rise in factory farming:** not just of livestock and birds, but also salad vegetables picked, priced and packed in industrial units (**adaptation**).

4 **Year-round crops:** a demand for the supply of certain crops both in and out of season has increased supply from foreign, usually cheaper, suppliers, for example Kenyan runner beans and Spanish tomatoes (**adaptation**).

Figure 29 illustrates how much employment in rural areas altered between 2008 and 2012 (**difference**).

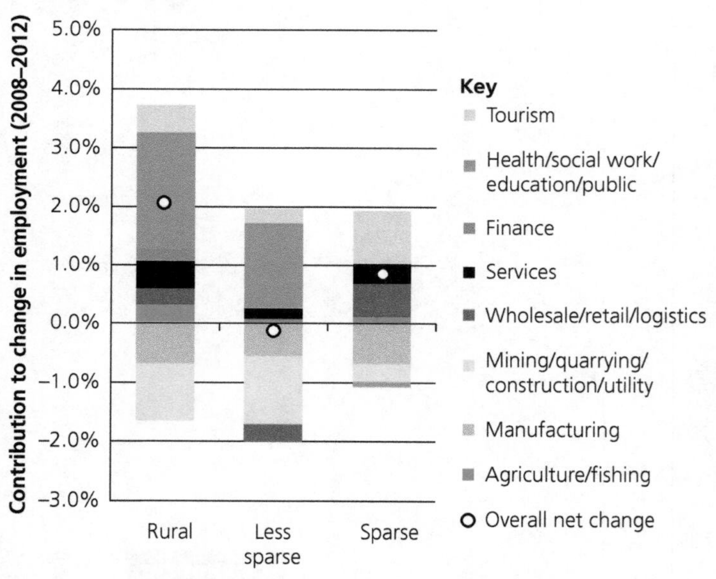

Figure 29 Employment change in rural areas of England (2008–2012)

Post-productive countryside refers to rural areas that no longer rely on the primary sector as the main source of income. Many such areas face a number of economic and social challenges (Table 21).

Knowledge check 18

Describe the changes taking place in rural areas. Are there any common elements? Identify the sectors that are growing and those that are declining in Figure 29.

Table 21 Challenges faced by the post-productive countryside

Challenge	Details
Agricultural change	Impact of mechanisation, foreign competition, reliance on foreign labour, climate change
Decline in services	Lack of economic viability has resulted in public and private services decreasing in communities
Lack of affordable housing	The growth of villages as commuter settlements and purchase of second homes by an affluent urban-based population can gentrify rural areas, pushing up house prices and limiting availability, excluding lower-waged rural workers
Depopulation	The young and the less affluent move away for better job prospects or increased opportunities of home ownership
Poor transport infrastructure	Limited or non-existent public transport can limit the use of other services; use of a car is expensive, yet in very rural areas travel by car is 20% higher than in urban areas, travelling 50% further
Access to superfast/ultrafast broadband	Rural broadband speeds are likely to be slower than urban areas, which may hinder the growth of new businesses
Fuel poverty	4% more households are fuel poor in villages than in urban areas; often rural houses are less energy efficient and rely on more expensive types of fuel for heating

Exam tip

Primary economic activity includes forestry, fishing and quarrying. Be sure to be ready with examples of the reasons for the loss of jobs in these industries.

Diversification in the countryside through reimaging and regenerating rural places

Rebranding is the way a place is changed and marketed so that the image and perception of it are improved. Rural diversification can be achieved by:

- **regeneration** — a long-term process aimed at improving the economy and social environment of the area
- **reimaging** — changing the reputation and perception of the area by the use of specific improvements

Exam tip

If using a case study of a rural area do not just write about 'change'. Be clear if the change is due to rebranding, regeneration, reimaging or a combination of more than one.

The rural idyll

Rural areas have been idealised in many people's representations. People tend to focus on rural natural landscapes that they wish to conserve, due to inherent beauty/ruggedness or because they live there. Rural areas are where our food is produced, for the big supermarkets and organic, local and sustainable small-scale retailers. Rural living has a 'gloss' that tends to hide socially excluded groups, such as travellers. Figure 30 schematises the rural idyll and how it is perceived (**identity**).

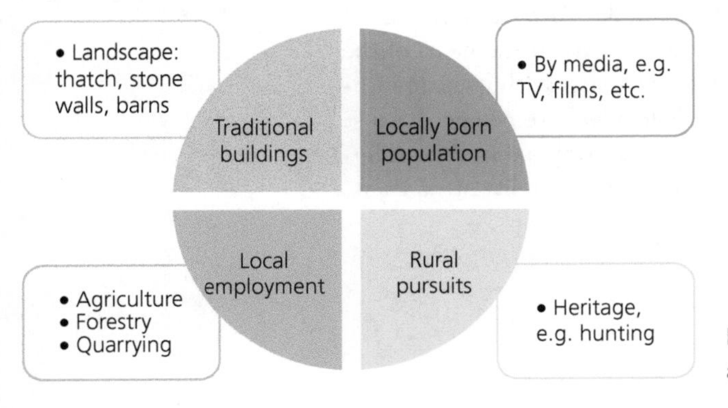

Figure 30 The rural idyll and how it is perceived

The perception of rural areas is frequently influenced by what is seen in the media, most notably on television and in film. This frequently presents a nostalgic version of the countryside. Television programmes such as *Downton Abbey*, *The Vicar of Dibley*, *Doc Martin*, *Hidden Villages* and *Escape to the Country* all reinforce perceptions of rural areas (**identity**).

Strategies for the rebranding of rural areas can be based on a number of themes (Table 22).

Table 22 Rebranding themes for rural areas

Theme	Explanation	Examples
Recreation	Increasing access to areas for leisure activities such as walking or cycling, or more adventurous activities	Swale Trail in Swaledale, Yorkshire Dales — a 20 km family mountain bike trail Zip Trekking Adventure in Grizedale Forest, Lake District — a 3 km network of seven tandem zip wires
Heritage	Encouraging an interest in historic buildings or the rural/industrial past	Amberley Museum and Heritage Centre, West Sussex — an industrial heritage site located in a former chalk quarry St Fagans National History Museum, near Cardiff — over 40 re-erected buildings set in the grounds of a castle
Media	Use of a location in a television series or film to attract visitors	The television series *Outlander* (2014–2018) was popular in the USA and increased numbers visiting Scotland Increase number of visitors to Northern Ireland to see locations used in *Game of Thrones*.
Event management	Using a location to hold an event or festival	Goodwood Festival of Speed and Goodwood Revival attract up to 200,000 visitors to the Goodwood Estate, West Sussex, each year Latitude Festival — music festival held at Henham Park, Suffolk
Food and produce	Advertising the range or specialism of foods from the area	Welcome to Yorkshire's creation of food and drink trails Big Feastival — a food and music festival in the Cotswolds

Rebranding strategies can be initiated by both formal and informal agencies, ranging from small groups of volunteers to local authorities, tourist organisations and external agencies, such as destination marketing organisations.

Knowledge check 19

What might be the disadvantages of rebranding based on a television series?

Local initiative reimaging — Cartmel

Cartmel (Figure 31) is a village of 1,500 people close to the Lake District National Park, which has reinvented itself as a place with a distinct identity. At the heart of the village is the twelfth-century Priory Church and the village square, with its sixteenth–eighteenth-century buildings. There is a small traditional steeplechase racecourse just outside of the village. The village website states: 'Cartmel is not trapped in time. The village offers today's visitors many modern surprises in the form of quality attractions, shops, eateries and places to stay.'

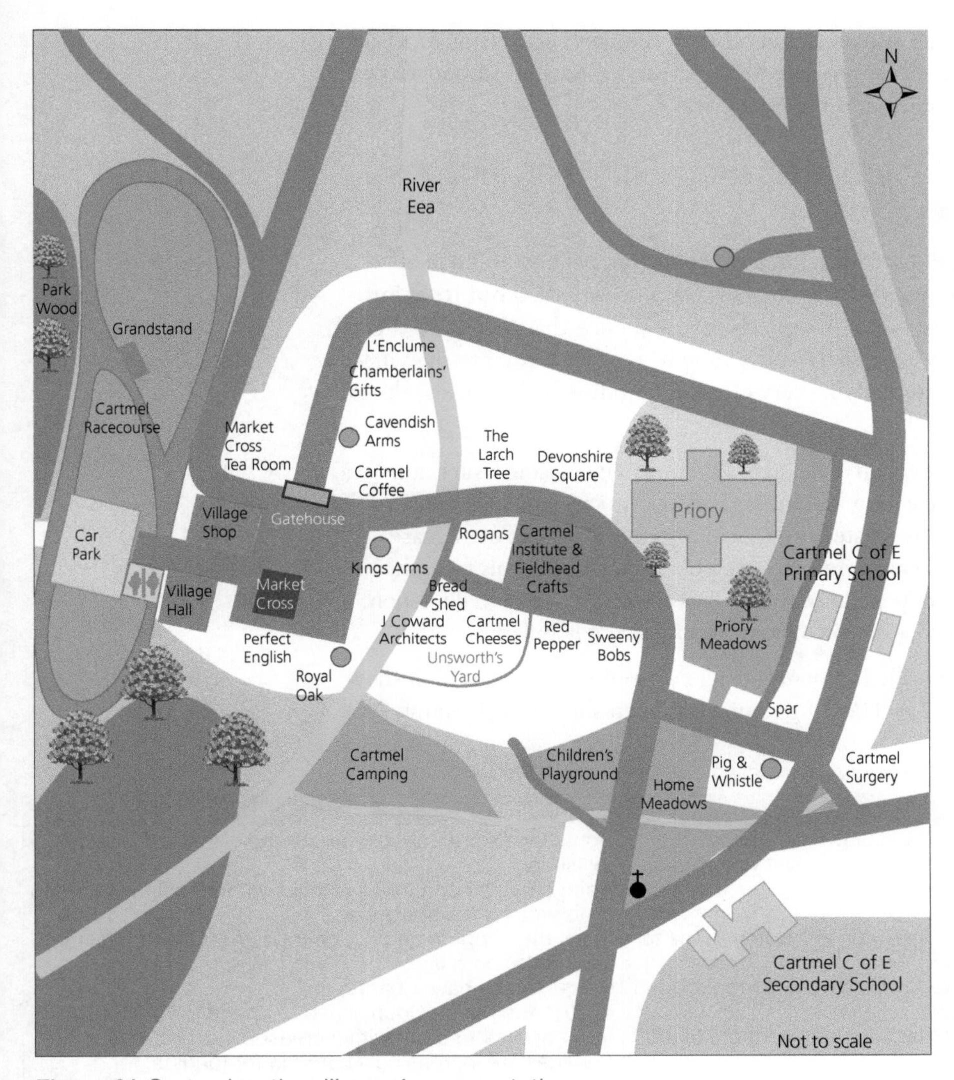

Figure 31 Cartmel — the villagers' representation

Reimaging in Cartmel can be traced back to 1993 (**time**, **risk**), when the new owners of the village shop took the initiative to close the post office and move into the marketing of souvenirs, including the Cartmel sticky toffee pudding. They had the foresight to patent the name, which provides further income and has aided the improvement of the village store. The puddings are now sold online and in shops such as Waitrose.

In 2002 (**time**) Simon Rogan, who had trained as a chef in various hotels in Hampshire, established L'Enclume in the old forge. In 2020, the two-star Michelin restaurant was named the number 1 restaurant in the UK by the *Good Food Guide*. Other food providers have located in the village to benefit from the growing gastronomic cluster, such as a cheese shop and a quality wine shop (**threshold**). Indirect publicity via Chris Evans when he was on Radio 2 has also resulted in more people discovering Cartmel and, in turn, more visitors travelling there (**media**). Even the racecourse is now the venue for stadium-style concerts in the summer months.

The popularity of the village has also attracted other businesses showcasing local artists and crafts. In 2014 the village produced the Cartmel Township Initiative (note that both 'town' and 'village' are used in the publicity): www.cartmelvillage.com

The consequences of rebranding on perceptions, actions and behaviour of people

Rebranding can have a range of outcomes for, and effects on, people. Some people are attached to the memories of a past 'golden age' — a process called **habituation** or **NIMBYism** ('not in my back yard') — with the consequence that they wish the village to remain as it once was (especially at the time when they moved to the village from urban areas, seeking the rural idyll). Very often, rebranding of a village seeks to play on past images.

Rebranding can improve the perception of an area, resulting in increasing visitor numbers, which benefits local businesses catering for tourists and may provide employment opportunities. However, non-tourist related services may go into decline. An increase in the demand for holiday homes can increase house prices, excluding locals from the market. As an area becomes more popular so there may be increased traffic congestion and environmental impact, which is resented by the local resident population.

Port Isaac in Cornwall has been the fictional setting Portwenn in the television show *Doc Martin* since 2004 (Table 23). The programme has been shown in 70 countries.

Table 23 Impacts of the television show *Doc Martin* on Port Isaac, Cornwall

Positive impacts	Negative impacts
Increasing visitor numbers — 27% of viewers who visit Cornwall and 43% of visitors to Port Isaac say it was due to the programme, with a rise in foreign visitors, especially from the USA Visit Cornwall states that the programme is worth millions to the Cornish economy Village appearance has improved, with cottages redecorated House prices have quadrupled The film company has made at least one payment of £50,000 to the village The Doc Martin Community Fund gets royalties from the show, which have been used to provide facilities for the local school and village hall The cast and crew stay locally and use local facilities for 3 months when filming	Some locals are unhappy about the disruption caused by filming Road closures and traffic disruption during filming create problems for residents Many day trippers and coach trips arrive but bring little income to the village House prices have quadrupled In 2011 the Parish Council had to spend £475,000 on a new car park to cope with increased visitor traffic Services are increasingly catering for tourists, for example the former site of Barclays Bank now sells Doc Martin souvenirs

Rebranding can also harness perceptions of a community and community spirit by encouraging volunteering in place of local government services.

Rebranding Blaenau Ffestiniog

Blaenau Ffestiniog, Wales was founded as a town based on primary industry (the quarrying of slate). Its peak population was 11,274 in 1881, but the decline of quarrying has led to a decline in population, which reached just 4,900 in 2011. In the late twentieth century, the townspeople had attempted to rebrand the town as a tourist centre (Ffestiniog Railway, Llechwedd Slate Caverns). But they wanted something more, especially a foothold in the growing adventure tourism industry.

Regeneration of the town has been a result of both private and public partnership:

- Blaenau Ymlaen was set up in 2006 as a local partnership between the community and the researchers developing the project. It was difficult to gain funding due to the scars of large-scale quarrying (**rebranding**).
- Local authority Gwynedd County Council, together with the Welsh Government and European Union funding, worked to improve public areas such as the town centre (**external funding**).
- The town was boosted by receiving awards from the Royal Town Planning Institute and the Institute of Civil Engineers Wales, and by becoming the Towns Alive Environment and Culture winner in 2013 (**actions of people**).
- Antur Stiniog 2007 was a locally developed community initiative that established mountain bike trails in the region. Today, Antur employs 19 people who administer the trails together with an outdoor equipment shop in the town centre. It also runs a fell-running competition, walking, kayaking, caving, and nature and history trails (**changing businesses**, **tourism**).
- The 2015 Velorail was established, using sustainable bike technology, along a disused railway line (**recreation**).
- Zip World Titan was built in the old quarries, to the point that its spatial extent is greater than anywhere else in Europe (**recreation**, **tourism**). Zip World has created 450 jobs in 5 years. The average visitor to Zip World spends £250–500, which is higher than the UK average visitor spend of £160.
- Bounce Below (giant trampolines and slides) was assembled in a former slate mine (**recreation**, **tourism**).
- The project Y Dref Werdd ('Green Town') addresses sustainability by working to reduce social problems among the population, improving the health of the population and creating a community of stakeholders (**behaviour of people**) who are passionate about the environment and who contribute to community development. The project has developed new allotments and is introducing smart energy to the village (**local community**).

Knowledge check 20

In 2014 250,000 visitors came to Blaenau Ffestiniog. However, the older generation is sceptical about its success — why?

The Great British High Street competition

The Great British High Street competition awards villages and small towns for their efforts to reimage themselves. In 2015 West Kilbride in Ayrshire, Scotland won for transforming itself into a craft village, supporting local artists and turning around the village perception with creativity and enthusiasm (**local community**). In 2016 Pateley Bridge won for reducing the number of empty units to only one, when there were 12 previously, and for community spirit, its social media campaign and a commitment to supporting local retailers. See also page 77.

Summary

- Not all rural areas are peripheral areas.
- Employment in primary occupations in rural areas is declining.
- Employment in rural areas is diversifying.
- Rural places are reimaging themselves either with governmental assistance or through private initiatives.
- Not everyone may be in favour of rebranding due to its impacts.

Rural management and the challenges of continuity and change

Processes affecting rural areas

Managing rural change

Since the 1980s the rural population has increased as a result of **counterurbanisation** — the movement of jobs and people out from cities to smaller towns and rural areas. This movement began with the relocation of manufacturing to the New Towns, outside of major cities, but the areas are now dominated by tertiary and quaternary jobs.

People have also migrated to gain the perceived advantages of rural life, while retaining the attributes of urban living, such as urban media, telecommunications and transport availability. Nevertheless, rural areas are dominated by the over-55s and the young still migrate to the cities to study and/or work. 25% of the 630,000 people living in Welsh rural areas are over 65 years old.

Demographic

Rural population is changing both in its demographic characteristics and location. The rural population had an average age of 41.5 in 2002, which increased to 45 by 2016 — 5.5 years older than the average for urban populations. While the average age for England rose by 1.1 years between 2002 and 2016, in most rural areas it rose by 3.1 years. Figure 32 compares the age profiles for 2001 and 2017. One-person (age 65+) households formed 14% of rural households and other single person households (13%) in 2011. Many were returnees to their place of birth, which has been estimated to be as high as 18% of all movement into rural areas in parts of Wales. Rural populations are generally healthier, despite the age of the population, and life expectancy is slightly higher than in urban populations. By 2025 it is estimated that the rural population could increase by 6%.

Exam tip

A population pyramid is a graph showing the distribution of various age groups in a population. Make sure you understand what future implications will be for a place based on the current shape of the graph. An internet search may find you the population pyramid for your home place which, if relevant, you could contrast with a rural area. You can also find how the populations of places have changed over time.

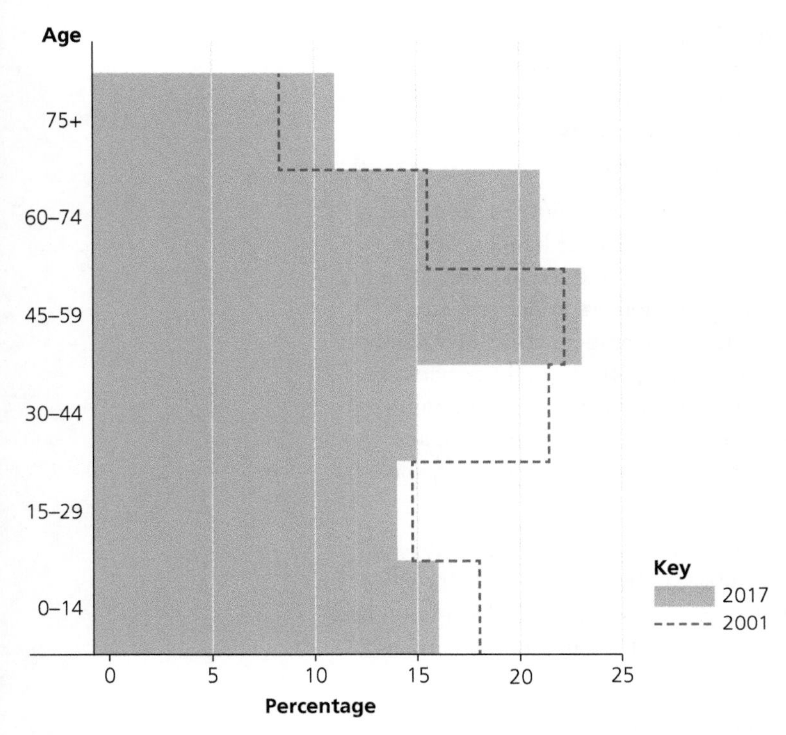

Figure 32 The age of the resident population in rural areas 2001 and 2017

Counterurbanisation

Migration of higher- and middle-income groups from urban areas seeking affordable property has swelled rural numbers. Figure 33 gives data on occupations of rural residents in 2011. It is notable that approximately one-third of those employed are in high-status, high-salaried jobs.

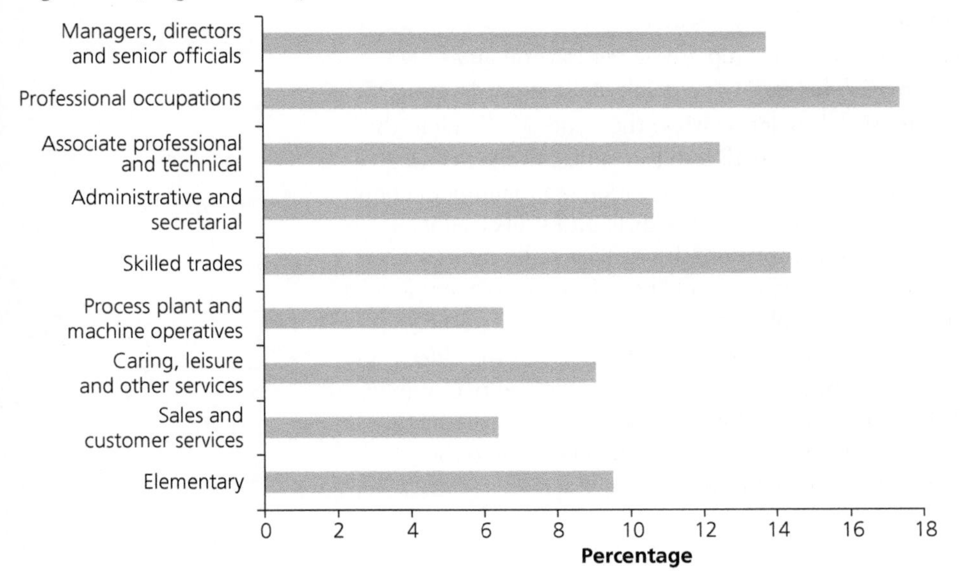

Figure 33 Occupations of rural residents aged 16–74 in England and Wales 2011

Knowledge check 21

How does the information shown in Figures 32 and 33 help explain some of the issues faced by rural areas?

All too often counterurbanisation is portrayed as individuals escaping the polluted, harried city for a more down-to-earth way of life in the country. Some of what is often called counterurbanisation is actually rural-to-rural migration, which studies in Wales and Scotland have shown to account for one-quarter of migrants. Many of these migrants have higher incomes than those leaving urban areas.

The traditional counterurbanising people are those able to commute to urban areas for work and shopping, and subsequently retire in their village or hamlet. This is a fast-growing characteristic of the rural areas fringing cities and towns. Areas fringing cities in southern England up to 130–180 km from London are under pressure to expand. Here, rural settlements are in high-quality countryside, within easy reach of rail and motorways, and have inhabitants whose work patterns and lifestyles are urban. Affluent rural areas (the home of 'Motorway Man') surround the growing urban areas of the south. These are the classic sites for counterurbanisation and give rise to pressures on land and housing for the diminishing number of primary industry employees and young people.

In contrast, in parts of the north of England and in parts of Wales, motorways and rural lifestyle have attracted people into the countryside from towns and cities suffering from deindustrialisation and multiple deprivation. Counterurbanisation here is from deprived towns such as Burnley, Preston and Bradford to affluent rural areas such as Ribble Valley, Wyre and Craven.

Counterurbanisation has the following effects on the area:

- **Lower-income local buyers:** local buyers can be priced out and the social mix of places altered as a result.
- **Second-home ownership:** second-home owners take properties out of the local market, which affects all areas, especially more sparsely populated districts. In Cornwall the takeovers have been so extensive that communities are looking at ways of alleviating the effects. In Scotland, ultra-high-net-worth individuals, often global investors, invest in Scottish rural estates and encourage a change in the nature of the rural economy towards leisure pursuits.
- **Owner occupation:** levels of home ownership are higher in rural areas (74%) compared with urban areas (61%). This is partly a result of counterurbanisation.
- **Changes in the socioeconomic characteristics:** those already living in the area are more likely to be involved in jobs with lower incomes, in the primary and care sectors. Low incomes lead to resource deprivation, an inability to obtain affordable housing and opportunity deprivation (the difficulties involved in accessing services such as health and recreation). Increasing skill requirements in all forms of primary activity leads to a greater chance of unemployment among those who lack qualifications. Rural people may experience in-work poverty.
- **Mobility deprivation:** this includes refusing driving licences for the elderly on the grounds of poor health, rising transport costs and poor availability of public transport as rural bus subsidies are cut, all of which leads to the inaccessibility of jobs and services. This may not been an issue when people move to the countryside, but occurs as they get older or as services decline.

- **Digital exclusion:** Figure 25 on page 57 identifies the areas in which digital exclusion is prevalent. You should be able to explain this in terms of peripherality, population densities and distance. As more services rely on the use of the internet, people who have moved to rural areas may find access to services increasingly difficult.
- **Poverty:** for some, poverty is one outcome. People are unable to share in the lifestyle of the majority in a small settlement because they lack the resources to do so.

Rural populations can be overly disadvantaged. Although the characteristics of deprivation are similar to those in urban areas, many have a disproportionate effect, which leads to a self-sustaining spiral of rural deprivation. Peripherality and isolation only make the effects of deprivation worse. Figure 34 summarises the forces that lead to reimaged rural places.

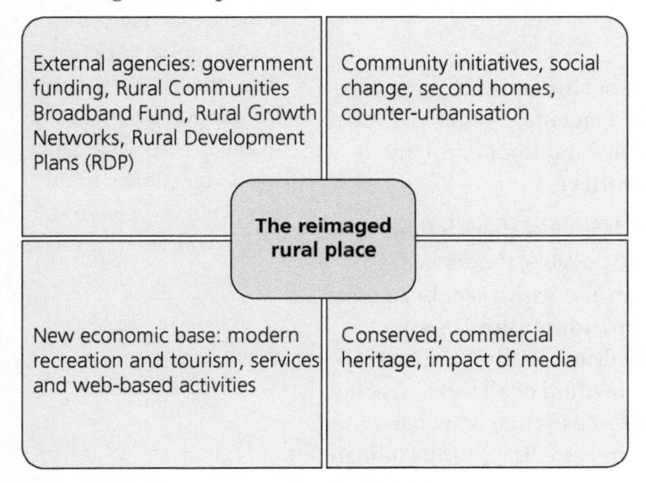

Figure 34 The reimaged rural place

Managing rural change and inequality in diverse communities

People tend to visualise places in their minds. The variety of people living in rural areas will inevitably result in different perceptions and understanding of the nature of that village or locality. In the mid-twentieth century Ceredigion and Montgomeryshire had a declining population. This coloured the perceptions of these places as being static and failing to move with the times. On the other hand, some saw this as an opportunity to acquire cheaper second homes. This in turn raised opposition from the local population and resulted in the attempted burning of some second homes. Infrastructure, access to schools and health facilities were poor, but these issues did not concern the second-home owners because they could return to the cities (**identity**).

Second-home ownership

Over 165,000 people own holiday homes in England and Wales. The greatest number of second homes is in Cornwall (23,000), whereas Gwynedd (7,784) has more per head of the resident population. In the Lake District 15% of properties have owners who live elsewhere, while in north Norfolk and South Hams, Devon the figure is 9%. For some settlements the figure is much higher. Around Padstow in Cornwall 67% of properties are bought as second homes and at Beadnell in Northumberland the figure is 50%. This tends to result in pushing up the prices of houses and long-term rentals in the most popular tourist regions, all out of reach of the local workers on low wages. In 2018 80% of locals in St Ives, Cornwall voted in favour of newcomers not being allowed to buy new-build developments.

Transport, health and service provision

The lack of transport in rural areas is a cause of growing inequality. People travel further than urban dwellers and mainly by car (59%), but 11% do not have access to a car (compared with 28% in urban areas). This reflects the relative wealth of the new rural population and the relative poverty of the long-term residents. Surveys have shown that one-third of rural residents find public transport inadequate because of irregular service and problems fitting in with working hours, school and service provision times (**isolation**).

As the proportion of elderly residents rises, supporting the health needs of the people living in rural areas is an increasing challenge. The percentage of people with disabilities or health problems that limit activity is higher in rural than in urban areas. This is particularly so in the more remote areas (**peripherality**).

Service provision can be under threat in some communities. With few shops left in villages, the closure of a post office is seen as a threat. Where possible, the service is moved and combined with another use, but there has to be an alternative site. Businesses no longer wish to be associated with a post office (**mitigation**, **adaptation**). Some chains have opened village stores, but their profits leave the village and the effect on local spending is small. Banks have also been closing, with one-third of all local branches closing in the last decade, many of them in rural areas. In 2014, 243 rural branches closed as a result of the success of online banking. Cash machines are also disappearing in many areas as more people go cashless. However, poor broadband speeds in more remote areas have had an impact on communities, where access to broadband is limited (**isolation**).

On the other hand, rural-to-rural migrants remain in rural occupations, sometimes branching out into food and accommodation businesses, which can employ others. Those coming from the cities and towns are more likely to set up arts and crafts and IT-related businesses, which rarely employ extra staff. Younger migrants into the country earn more and are in higher professional occupations. The reasons for moving to or between rural areas are shown in Table 24.

Table 24 Motivations for moving to rural areas

Motivation	Counterurban migrants	Rural-to-rural migrants
Employment	11%	22%
Housing	3%	10%
Nearer parents/adult children	10%	10%
Retirement (planning for, or actual)	33%	28%
Quality of life	26%	15%
Other	17%	14%

Source: Stockdale, A. (2015) 'Contemporary and 'messy' rural in-migration processes: comparing counter-urban and lateral rural migration', in *Population, Place and Space* (John Wiley & Sons).

Exam tip

Be wary of your data sources and whether they may be biased. The figures given in the text for second-home ownership in districts do not mention whether the owners are joint rather than individual owners of the properties, or if the properties are used as holiday lets or left empty for part of the year. Also, the St Ives reporting initially implied that even existing second homes would be subject to the ban when they were sold, which was later shown to be false.

Ongoing challenges in rural places where regeneration/rebranding is absent

Some of the greatest challenges face rural settlements that are too small to be the subject of a major regeneration scheme. The challenges of a small population threshold are being addressed by minor changes that may impact on the community. For example:

- In the UK there are over 350 community-run shops and/or post offices, some staffed by volunteers. An estimated 300–500 village shops close every year. Since 2010 an average of 22 shops have opened under community ownership each year. Pwllglas Community Shop, near Ruthin, Barford Village Shop in Warwickshire and St Tudy Community Shop and Post Office in Cornwall are examples of this trend.
- Forncett St Peter, Norfolk uses an old telephone box as a site in which to house a defibrillator, which caters for health emergencies in the remote village.
- The Upper Dales Community Partnership based at Hawes in the Yorkshire Dales runs a number of services that would have disappeared from the area. Using some part-time and around 60 volunteer drivers, it runs the Little White Bus service, which replaced the local bus service when it ceased to run in 2014. Carrying over 60,000 passengers a year, the surplus money helps to subsidise the post office, which it also runs from a community office that is also the local library and council enquiry office. When the local petrol station closed the partnership opened the first community-run filling station in the UK. It also manages a small business park and is building some affordable homes for renting.

As well as experiencing a lack of service provision, such areas may continue to suffer from out-migration, especially of the younger population in search of employment. The lack of superfast broadband may limit investment in the area and hinder business start-ups.

Exam tip

Be careful of making sweeping generalisations. Just because a rural area has not been rebranded it does not mean it must be in decline with only an unskilled, elderly population. Many rural areas contain an economically active community with highly skilled residents.

New challenges of managing change associated with counterurbanisation and second-home ownership

Community Land Trusts

Community Land Trusts (CLTs) are one strategy being used to counteract the pressure of counterurbanisation and second homes. CLTs were first used in the USA during the Civil Rights Movement of the 1960s, and in areas of urban housing shortage. The principle was brought to the UK and the National CLT Charity was launched in 2014 as a response to the early success of CLTs. CLTs are themselves a response to the loss of services (shops, post offices, doctors' surgeries) and high house prices as counterurbanisation changes the settlement (**adaptation**, **mitigation**, **risk**, **sustainability**). They are run by the local community to manage homes and other assets in the community. They act as long-term stewards of housing, ensuring that it remains affordable, based on what people earn in the area. CLTs have to be constituted so that they can buy, sell and rent property, and be eligible to obtain loans. Resale of owned properties is restricted to 30% of the market value. For more details on CLTs, go to www.communitylandtrusts.org.uk.

St Minver CLT in north Cornwall was established in 2006 due to difficulties local people were having in affording homes. The area includes the village of Rock, a desirable holiday destination attracting affluent visitors and a second-home hotspot, reputed to be one of the most expensive places to buy a property.

The CLT has provided 20 self-build homes in the parish for local people, helped by an interest-free loan of £540,000 from the local authority. If the self-builders were to sell their home the CLT would have first refusal to buy the property or to nominate the purchaser, keeping the property in the hands of the community. The CLT is now looking at another phase of building to provide affordable homes.

Policies to restrict second homes

There are a number of potential policies that might be adopted in an attempt to limit the number of second homes in an area.

- Introduce a policy that requires all new residential dwellings to be restricted to a person's 'principal residence'. The Yorkshire Dales National Park has a restricted occupancy policy, which means that nearly all new homes will be reserved for Dales residents.
- Removing council tax subsidy on second homes, or increasing council tax.
- Additional stamp duty tax on second-home purchases.
- Providing financial assistance to first-time buyers in rural areas.
- Placing quotas on second-home building (used in the canton of Valais, Switzerland).
- Designating properties that can be used for holiday lets and second homes rather than permitting all to be available for these purposes.
- Withdrawing capital gains tax concessions on the sale of these properties.

Foreign direct investment

Foreign direct investment is also being used to regenerate some rural areas in which there has been a loss of primary activity. Investors from the USA, in partnership with a UK company, are proposing to transform a former open-cast mining site near Chesterfield, Derbyshire into a health, sport and education facility, which should create 1,000 new jobs.

Summary

- The processes of change in rural areas include those associated with demographics, counterurbanisation and the housing market.
- Population is growing in most rural areas due to counterurbanisation and return migration to rural areas.
- Movement to rural areas is often on the basis of the perceptions of these places — the rural idyll.
- An outcome of change is rural deprivation and poverty, which is very dispersed and may not be revealed by local area statistics.
- Second homes and holiday homes put pressure on the housing market.
- Efforts are being made at local governmental and community levels to address the problems of deprivation, the provision of services, housing and isolation in rural areas.

The rebranding process and players in urban places

Towns and cities compete as places to attract investment. In this urge to compete, those in power project images of a place that they know will attract the investment that they want. Canary Wharf has been successful not only because it benefited from Urban Development Corporation financial support from the government, but also because the place reflects the images of the finance industry — high-rise buildings, with well-dressed and well-qualified people commuting to work in a district that contains restaurants, bars, shops, hotels and apartments to cater for the high-earners. This is in marked contrast to the working-class and immigrant population — employed in the docks and living in terraced housing with a lack of infrastructure — that had characterised the place before 1980. The location is the same, but the *place* has changed; the Isle of Dogs has been rebranded as Canary Wharf.

Reimaging and regenerating urban places

Urban areas can attempt to change their image in a number of ways (Table 25).

Table 25 Methods of reimaging urban areas

Reimaging method	Description
Promoting business advantages	Advertising global links, suitable buildings and quality of life for employees
Development of stadia	Improvements are usually linked to a major sporting or cultural event, and act as a catalyst for regeneration
Promoting cultural quarters	Advertising museums/theatres to increase visitor numbers; promoting parts of a city that contain a distinctive cultural background as an attraction
Festivals	Promoting cultural events, food festivals or Christmas markets to increase visitor numbers
Industrial heritage	Restoring industrial areas as visitor attractions or as locations for small craft industries or specialist outlets
Flagship development	Promoting a high-profile land and property development to provide a catalyst for further development

Exam tip

Rebranding may only occur in part of a city. When using examples, be specific and give the name of the area rather than just the name of the town or city. However, to avoid confusion you should include the urban area in which it is found.

Reimaging and rebranding is sometimes called **boosterism**. The way a place is 'boosted' can change over time. Chicago was once called 'Gem of the Prairies' before it received negative publicity as 'Hogopolis' and 'Cornopolis', while today 'the Windy City' is its more benign tagline. Cities now promote both business opportunities and lifestyle opportunities, as shown in Table 26.

Table 26 Types of city promotion

Business promotion	Lifestyle promotion
Centrality Accessibility Communication costs Landscapes of modernity — clean, flagship buildings Specialist and quality skills — HE institutions Global links — airports, international train stations	'Most liveable city' Cultural centrality Centre of action — clubs, bars, theatres, sport Leisure time and refined leisure facilities Access to countryside

Sports stadia

Stadia are frequently used as catalysts for regeneration, although the trigger is often the result of the global pressure of hosting major sporting events, as the following examples demonstrate:

- Don Valley Stadium, Sheffield (1991) — World Student Games
- Principality Stadium (originally Millennium Stadium), Cardiff (1999) — Rugby World Cup, city centre site
- City of Manchester Stadium (now Etihad Stadium) — built to host the Commonwealth Games (2002); 1.6 km from the city centre on a brownfield site and part of Manchester Sportcity
- Olympic Stadium (now called London Stadium) and Park (2012), East London

The costs and benefits of stadia developments are summarised in Table 27.

Table 27 Costs and benefits of major stadia developments

Benefits	Costs/disadvantages
Can underpin regeneration goals Generate jobs Increased commercial activity Multiplier effect of the benefits above Bring tax income to local council and country Create landmark sites that identify the place/city Increased tourism (e.g. stadium tours, visiting fans, hotels) Increased community provision Generate civic pride Image improvement Raise sponsors' profile (e.g. Amex, Brighton; Etihad, Manchester) Revive property prices (e.g. Cardiff) Catalyst for property renovation (e.g. Cardiff)	Congestion Accessibility required for large numbers (e.g. Wembley) Noise and light pollution Quality of life issues for residents (i.e. NIMBYism) Property values may decline High maintenance costs Shortfalls in income from occasional use Conversion costs (e.g. Olympic Park) Policing on match days Vandalism, graffiti, litter

Other stadia have been relocated to enable the former site to be used for other activities. West Ham United's move to the London Stadium in 2016 has allowed the old ground to be redeveloped to build 850 new homes. Stadia are invariably multi-purpose and are used as venues for pop/rock concerts, community activities and conferences, which make use of the building's administrative space.

Culture-led regeneration

Culture-led regeneration was initially inspired by the success of the Guggenheim Museum in Bilbao, Spain. It is one method of promoting economic regeneration. European City of Culture designation commenced in 1985 and, so far, two UK cities have been designated — Glasgow in 1990, with its strapline 'Glasgow's miles better', and Liverpool in 2008. In 2009, a UK City of Culture distinction was established, with Londonderry taking the mantle in 2013, Hull in 2017 and Coventry in 2021.

Gateshead Quayside in Newcastle-upon-Tyne, including the Millennium Bridge, the Baltic Centre of Contemporary Art (formerly the Baltic Flour Mill) and the Sage Gateshead concert venue, is another prime example of regeneration, at a cost of £142m. Some jobs elsewhere in Newcastle could possibly be attributed to this development, but it seems to have had little effect on job creation in the immediate area. Rather, it has had other impacts in terms of the actual and perceived images of the city (**identity**).

In Bradford, the establishment of the Museum of Photography, Film and Television in 1983 and its achieving the first UNESCO City of Film status in 2009 have been pivotal in allowing regeneration to take place. In addition, the Moorside Mills (1875) that once wove worsted and lay derelict since 1970 were converted into an Industrial Museum, opening in 1975.

The Turner Contemporary Gallery, which opened in 2011, brought an extra £41m spending to Margate in the following 5 years as the result of approximately 200,000 extra visitors to the town. This resulted in a boutique hotel opening in 2013 in a converted old seafront hotel, and many more restaurants and arts-related retailing.

Multifunctional redevelopments that contain theatres as well as retailing, cinemas and even local government facilities have been favoured, but these tend to drain people away from the traditional high street. Theatre restoration can also be part of cultural regeneration.

The promotion of a cultural event can increase visitor numbers and help provide an urban area with an identity. The Edinburgh International festival is ranked as one of the most important cultural celebrations in the world. Combined, all the festivals held in Edinburgh attract over 4 million people.

Farmers' markets and Christmas markets are other forms of cultural regeneration designed to return shoppers to town centres. Christmas markets are frequently identified with the German model in an attempt to provide an image that is associated with affluence and possibly mitigate the declining status of a CBD's retail district. The annual Christmas market in Bury St Edmunds in Suffolk attracts over 120,000 people to the 4-day event. The event increases trade in the town, is seen as putting Bury St Edmunds 'on the map' and results in many visitors returning to the area at other times of the year.

Industrial heritage

Digbeth, Birmingham was home to the original Bird's Custard factory. In 1993 the factory was restored and now houses over 100 businesses — mainly creative enterprises who were attracted by the city centre location and the creative environment, which has its own sense of identity. Over 1,000 jobs have been created and regeneration has spread to the surrounding area.

Flagship developments

Flagship developments aim to encourage further regeneration of an area. The Stephenson Quarter in Newcastle aimed to regenerate a large area that had been underdeveloped for years, to help attract increased visitor numbers and spending within the city. The development includes residential, hotel, conference, commercial office and leisure uses, as well as car parking, exhibition and events venues and retail facilities.

Reimaging and regenerating urban places through external agencies

Rebranding requires places to shed their old image and to metamorphose into new, reimaged places that use the past as a part of the new brand image (**adaptation**, **identity**, **representation**). Different agencies can be involved in urban rebranding, ranging from governments to community groups.

Local authority government rebranding

In 2017 there were 227 active **Business Improvement Districts (BIDs)**, the majority in urban areas. A BID is a partnership between a local authority and local businesses to provide additional services or improvements to an area. The BID is funded primarily by a levy on the local businesses in the area.

Newcastle NE1, started in 2009, is widely recognised as one of the most successful BIDs. The Newcastle initiatives targeted the needs of city centre retail businesses after 5 p.m., aiming to:

- extend shop opening hours, with free parking
- build on the number of extra visitors to the Eldon Square shopping centre that followed the extended opening times
- improve access to the Central Station
- use vacant shops as spaces for youth training, employment support, entertainment and socialising (Space 2)

By 2019 the BID had created an evening economy worth £839m. It secured funding of £25m for the redevelopment of Central Station, which created 166 direct and 1,679 indirect jobs. It has secured a further £3.2m to redevelop the Bigg Market area, which should attract £60m of private investment in the area. It established the Newcastle City Marina, which paid for itself five times over, with visitors adding £250,000 to the economy. A youth employment programme is in operation. There have been an additional 13.7 million post-5 p.m. visitors to Eldon Square shopping centre since the scheme launched. Events organised or supported, such as restaurant week, attracted 340,000 visitors in 2017, adding millions in economic impact.

BIDs in smaller settlements are less successful because retailers and services are expected to contribute to the scheme. In Skipton some services, such as hairdressers, resent contributing to a scheme designed to attract tourists.

Tramshed, in Grangetown, Cardiff is an example of a regeneration project based within a Grade II listed building that was formerly a tram depot. It has become a 1,000-capacity live music venue with a restaurant, cocktail bar and small cinema, an art gallery, dance studios and some office space.

It pays to be wary of job creation figures in regeneration projects because most overestimate job creation by up to 40%.

Regeneration by corporate bodies and community groups

Banks, alternative lenders, crowd funding, big companies and universities are all involved in backing new economic activities in cities. StartUp Britain is a national campaign launched in 2011, founded by eight British entrepreneurs to encourage enterprise in the UK. Since 2014 it has been part of the New Entrepreneurs Foundation charity. With the backing of businesses, it offers support and advice to new enterprises. Its website www.startupbritain.org has interactive maps that plot the density and number of new businesses. LEPs (page 36) are also involved, such as in the Innovation Birmingham scheme on the 14 ha Aston University Science Park.

Stalled Spaces Glasgow

Glasgow currently has more vacant sites than all other Scottish cities. Site plans for future development have been completed, but it can be up to 10 years before development is scheduled to start. Stalled Spaces Glasgow is a programme introduced by Glasgow City Council to support community groups and local organisations in developing temporary projects on stalled sites or under-utilised open spaces, with the aim of improving community health and wellbeing. 25 ha of stalled space have been used for: a green gym/play space, outdoor education, arts projects and pop-up sculptures, an urban beach and exhibitions. In this case, place and space are being used to improve lives in areas where there are environmental and socioeconomic constraints on health and wellbeing.

In Wales, the **Vibrant and Viable Places Fund 2013** provides Welsh government money for regeneration in Bridgend, Colwyn Bay, Deeside, Holyhead, Merthyr Tydfil, Port Talbot, Newport, Pontypridd, Swansea, Pontypool and Wrexham. In addition, these funds have been granted to tackle poverty in Tredegar, Rhymney, Grangetown, Cardiff, Llanelli, Rhyl, Caernarfon and Barry.

Town Centre First and Great British High Street

Much city centre regeneration has focused on retailing. Town Centre First policies, initiated in 2013, are an attempt to redirect new retailing away from out-of-town sites. They require that funds are used for the regeneration of areas within the town centre. Funds may be used for developments on the edge of town centres if suitable central sites are not available. However, concentrating on the centre has been shown to have a negative effect on small corner shops and district shopping centres, besides not always halting out-of-town developments. Some high streets are small, badly located and have little to no chance of competing with the internet and out-of-town retail destinations. Consequently, the policy is of little to no benefit to such places. Regeneration is unlikely and some planners recommend that the decaying main street should be transformed into housing or office space.

In 2018 the annual Great British High Street competition (page 65) awarded a prize to Crickhowell in Wales as one of the last high streets free from major retailers, where every shop except one was independent and family run.

Flagship development: the case of King's Cross (private rebranding and reimaging)

The area around King's Cross and St Pancras train stations was sandwiched between rail tracks and marshalling yards, gas holders, derelict warehousing and industrial buildings, with contaminated ground, a canal and some nineteenth-century housing of poor quality. Nearby, the British Library was built and involved the eviction of 2,000 people from existing housing. Eurostar opened at St Pancras in 2007, yet the 27.1 ha area north of the two stations remained undeveloped following the granting of planning permission in 2006 until 2012, when Google announced that it would move to the site.

Figure 35 shows the site, the balance of planned land uses and their locations in a mix of old refurbished buildings (e.g. the Granary has been converted into the home of the newly created University of the Arts). The conserved shells of the gas holders have been moved to create space for new flats. 40% of the site will be open space.

The remainder will house 280,000 m^2 of workspace (mainly offices) and 46,500 m^2 of education, retail and leisure uses, together with 2,000 new homes and student residences. The whole scheme is due to be completed in 2021. By 2019 20 new streets and 50 new buildings had been constructed. The area included 1,900 new homes and there were 10 new public parks and squares, including one park inside the frame of a restored gas holder. There are 10.5 ha of open space and new retail areas, such as Coal Drops yard, which houses 50 independent shops and restaurants.

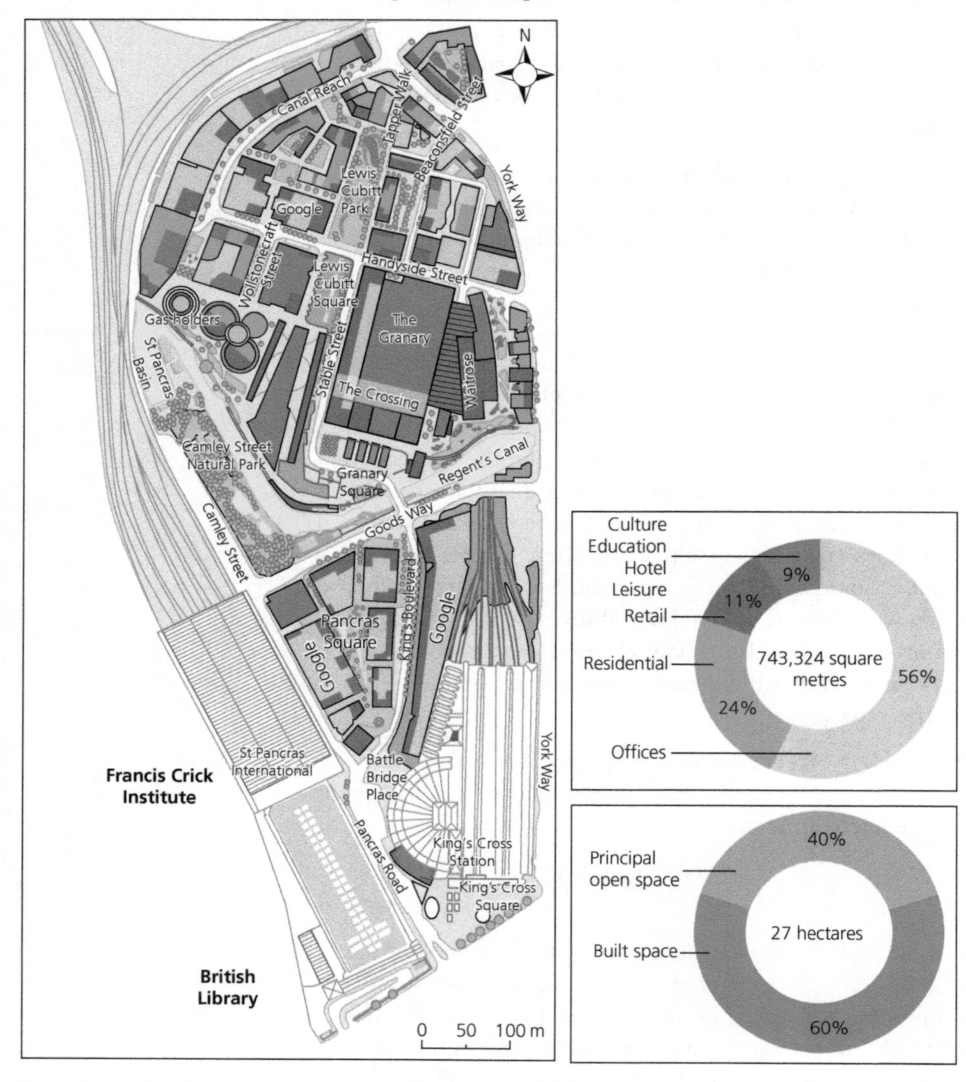

Figure 35 King's Cross regeneration area

The uniqueness of this redevelopment is that it is a private-sector development that has taken over former public-owned land. The area is being developed by the King's Cross Central Limited Partnership, which involves Argent and Hermes Investment Management on behalf of the BT Pension Scheme. The third partner is Australian Super, the biggest pension fund manager in Australia. Increasingly, large-scale redevelopment is being placed totally in the hands of private rather than public authorities, because it makes raising the necessary cash much easier (**difference**, **globalisation**, **representation**, **risk**, **sustainability**).

At www.kingscross.co.uk/discover-kings-cross you will find a variety of articles about the King's Cross project. This secondary information can act as a good example of an alternative place and a good point from which to re-examine your understanding of many of the key concepts. Some of the key concepts are addressed by this privately funded project, and by reading the articles you can use King's Cross to illustrate others:

- **Time:** the site has changed from industrial and transport dominance into a mixed-use site adapted for city living and working patterns in the twenty-first century. There is a long-term historic timescale of 150 years and, for the recent developments, a timescale of decades.
- **Place:** this portion of geographic space has uniqueness and distinctiveness as a result of the way it has developed and changed. The confined site may or may not relate to other places and spaces at a range of scales. Does it have identity? If so, who is determining that identity — developers, users, residents? Does it show that layered history?
- **Identity:** how do developers, users, residents and visitors experience the development? What does the place mean to them?
- **Globalisation:** the impact of global companies, such as Google, on the development. The international rail links and hotels are evidence of an increasingly interconnected world.

Heritage Lottery Fund (HLF)

The HLF opened in 1994 — www.hlf.org.uk gives a range of examples of the support for conservation, renovation and regeneration that the fund is providing in six fields:

- Land and natural heritage
- Museums, libraries and archives
- Buildings and monuments
- Culture and memories
- Industrial, maritime and transport
- Community heritage

In Wales, the Coastal Communities Fund for projects creating jobs in coastal areas is supported by the National Lottery Community Fund.

Many of the schemes supported by the HLF involve retaining the character and **identity** of places so that they are **sustainable**. The **risks** to buildings, communities and landscapes have been **mitigated**.

The impact of urban reimaging and regeneration

Many areas that have undergone regeneration have had a positive impact, improving the appearance of the location and increasing employment opportunities for individuals. This in turn has attracted people and businesses to locate in the area, such as Google locating in King's Cross. The regeneration can also boost the economy and attract more visitors. However, that growth may be at the expense of other areas as businesses change location or individuals change their shopping habits, causing a decline in the non-regenerated parts of the city. The opening of Liverpool 1, a large redevelopment along the River Mersey waterfront, created an increase in shop vacancy rates in other parts of the city.

Perspectives of groups about identity

By the time you, the reader of this Student Guide, are 60 years old, over 25% of people in the UK will be over 65. There will be more people over 60 than under 15. Are places considering the implications of this massive change? Some places are trying to involve all age groups, such as Edinburgh's City for All Ages, to ensure full social and economic inclusion of older people. Joondalup in Australia has a Growing Old Living Dangerously (GOLD) programme to create recreational townscapes for the elderly. The World Health Organization (WHO) has listed the key dimensions of a city that is age-friendly (Figure 36). These dimensions could form the elements of an investigation into the changing nature of a place.

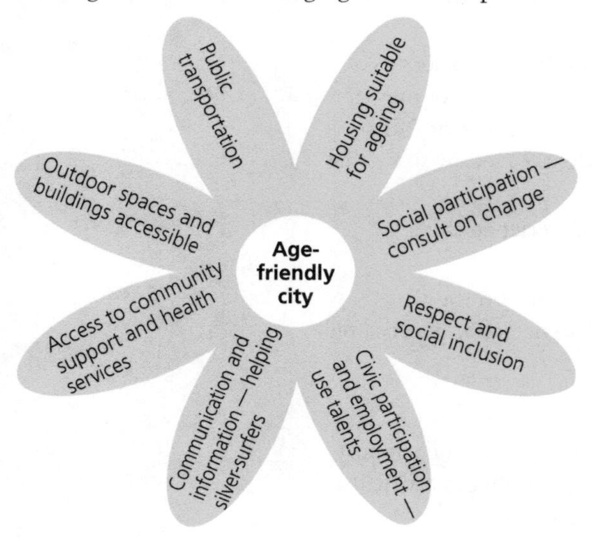

Figure 36 Aspects of age-friendly cities

For all of the policies and attempts to rebrand places, you need to ask 'who is gaining and who is losing'? This is a form of cost–benefit analysis based on welfare, and is often called **welfare geography**. Is a policy favouring: (a) one group of people, (b) businesses, (c) local inhabitants, (d) young adults, (e) the wealthy investor from home or overseas, (f) the elderly, (g) those on low incomes, (h) the homeless, (i) multinational companies, or is it potentially damaging the interests of these groups? Not all of those listed will gain or lose from a single policy intended to improve the physical and social environment of a place, but the impacts on the lives of some of these groups can be examined in any reimaged and regenerated area, whether it be St David's, Grangetown, Rotherham or the Olympic Park, East London. Leicester is one city that has recently been given a chance to reimage, through becoming Premiership champions (2016), being the home city of the former World Snooker champion (Mark Selby) and its cathedral becoming the site for the burial of King Richard III's remains.

Exam tip

Have your own examples of how the identity of a neighbourhood or a place has changed, preferably taken from a place that you know well.

Summary

- Rebranding and reimaging urban places is a continuous process although it has grown in importance in most developed countries.
- Promotion of places tends to be seen as a business-based promotional activity.
- Lifestyle reimaging is of growing importance and is often linked to the cultural regeneration of urban places.
- Local government has an important role to play in regeneration.
- Private enterprise involvement in rebranding is growing.
- The success and/or failure of all reimaging and rebranding must be judged on who gains and who loses.

Urban management and the challenges of continuity and change

The effects of reimaging and regeneration on the social and economic characteristics of urban places

Reimaging and **regeneration** are described by the Welsh government as:

> **...an integrated set of activities that seek to reverse economic, social, environmental and physical decline to achieve lasting improvement, in areas where market forces will not do this alone without some support from government.**

Where urban areas have undergone successful regeneration, it should change the perception of the area, such as making a place considered a safer and/or more attractive place to live. Many of the new employment opportunities are in the knowledge economy, which tends to attract younger employees. This, in turn, can lead to a higher-earning, younger population moving into regenerated areas and pricing out older, less-affluent members of society who may not have the required skills for the newly established employment. As well as jobs, the service sector attracted to regenerated areas may be aimed at the needs of the newer inhabitants, with the potential to exclude those who lived in the area before. Conflict with other areas may arise if they see a decline due to firms and people relocating.

Knowledge check 22

Why might some locals not be in favour of rebranding?

Cities are also being managed in other ways. Social media and technologies have transformed the ways in which we communicate, learn, work, consume, express emotions, relate to each other, and create and share information and knowledge. It is sometimes called a **ubiquitous commons**. Therefore, it is inevitable that the characteristics of places are perceived to be changing. Settlements have altered from being physical places on a site with links to other places, to places where data, information and knowledge are exchanged between people, while devices and data-storage points are held by organisations, companies and institutions.

Living in Safe Cities is the response of the Economist Intelligence Unit (EIU) to the challenges of modern urban life. This is a global study of 50 cities that has established the variables for deciding whether a city is secure. The index is the summary of four indices: digital security, health security, infrastructure safety and personal safety.

Smart Cities is a movement that envisages urban managers and technology companies working together to organise urban processes more efficiently, with the aim of improving quality of life. It is anticipated that IT in the form of artificial intelligence (AI), GPS tracking and the Internet of Things (IoT) can manage energy and water supply, transport, logistics, and air and environmental quality. It comprises six key fields, which are aggregated from 90 indicators in 27 domains. Globally it is estimated that over £1 trillion will be spent developing smart cities by 2030.

Bristol is Open is a joint venture between the University of Bristol and Bristol City Council, which helped Bristol to be recognised as the UK's leading smart city in 2017. Funded by local, national and European government, along with academic research funding and the private sector, it aims to provide a socially fair quality of life for all. The city is creating a digital network of 144 underground fibre cables, a mile-long stretch of wireless connectivity and 1,500 sensors on lampposts. This is combined with data sharing and analytics management to deliver services to public authorities, residents, businesses and visitors. The city aims to create 95,000 new jobs, especially in creative industries and green technologies, reduce carbon emissions by 40% by 2020 and have Bristol recognised as one of the top 20 European cities by 2020.

The **Transition Towns Network** was founded in 2005. It stresses the need for community-led change in response to rising energy prices and climate change. Transition towns utilise bottom-up initiatives to address food supply, transport, energy and housing. Examples include Brixton, which raised £130,000 to install the UK's first inner-city, community-owned power station consisting of solar panels on top of a council estate. Totnes, Devon has become one of the most advanced transition towns. Founded by community groups in 2011, Totnes Transition Network aimed to maximise local spending in the local economy (www.totnes.transitionnetwork.org). Of local spending on food, 66% was going to the supermarkets that bought globally. By encouraging local purchasing of food, three times more jobs could be created for the same spending on food. The local multiplier effect has been calculated to show that every £1 spent in local suppliers generates spending of £1.76 in the local economy, compared with only 36p if the same £1 was spent in a supermarket. Totnes also developed a local currency (the Totnes £) to encourage local spending but this ended in 2019 due to competition with online shopping.

Ongoing challenges in urban places

Challenges where regeneration/rebranding are absent or have failed

At the simplest level, urban areas with no regeneration may continue to decline. If businesses are attracted to other, newly regenerated areas the rate of decline can accelerate, possibly resulting in a cycle of decline and deprivation. Many urban areas are crowded and congested, making them unattractive for modern business without considerable investment. Likewise, urban areas remote from growth areas or lacking efficient communications may find regeneration difficult.

In 2014 Hall identified five twenty-first-century challenges for urban areas in the UK:

1 **Rebalancing our urban economies:** Figure 37 shows how some places have embraced the new economies while others lag behind.

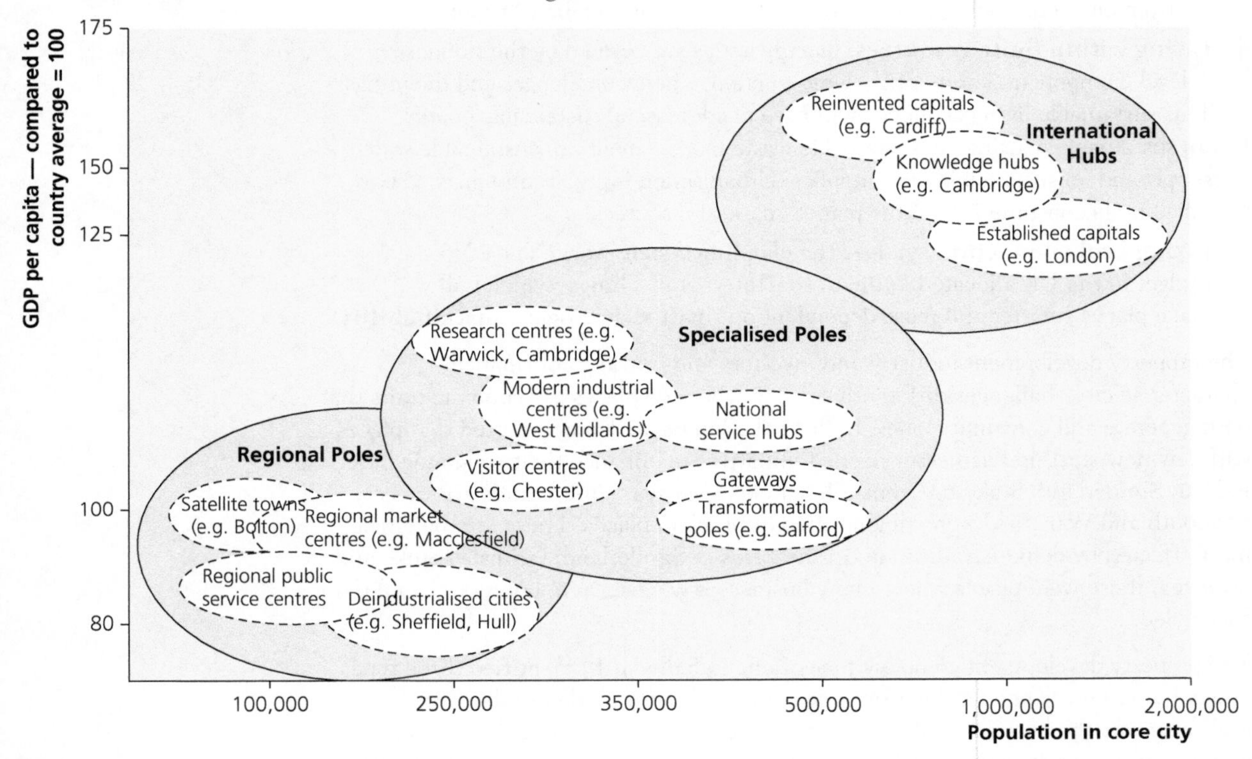

Figure 37 Embracing the new economies (Hall's categorisation of places)

2 **Building new homes:** where should new homes be built, and for whom? How do we secure brownfield land for homes? Bristol is under-supplied with homes. In 2013 the Bristol Homes Commission identified sites that best meet needs through freeing up land used by the council and by targeting 'greyfield' land, i.e. underused public amenity land, such as over-large car parks close to existing neighbourhoods.

3 **Linking people and places:** this challenge arises because many places have developed over a long time. Therefore, the public transport system linking people and workplace, often built over a century ago, needs updating to cater for the new work patterns and workplaces (e.g. Elizabeth Line/Crossrail). The Armitt Review 2013 examined the British transport infrastructure and recommended that transport planning must meet the needs of people and cover appropriate economic areas. The various authorities concerned with transport need to coordinate their activities, creating integrated transport authorities, as is the case in London (TfL).

   a Trams were abolished, but today new tramways are an efficient transport system in Manchester, Sheffield and Croydon. With a few exceptions, cities in the UK have not been built for the car or indeed the bicycle.

   b Adjusting roads to cope with modern patterns of workplaces and the volume of vehicles on the road is a challenge.

c Dutch cities and Copenhagen are among a few cities that have really embraced the bicycle. In the UK 43% of the lowest-income households and 66% of those on Job Seeker's Allowance do not have access to a car. If they are to benefit from city centre job growth there must be adequate public transport.

4 **Living within finite resources:** (energy issues are covered by the Eduqas A-level Component 3 and WJEC Unit 4 option, Energy challenges and dilemmas). This sustainable living challenge will have to address: (a) sustainable energy, (b) sustainable housing, (c) sustainable waste management, (d) sustainable water supply and (e) sustainable food supplies. Urbanisation is greedy and dirty. City inhabitants consume 75% of the planet's natural resources.

5 **Fixing broken machinery:** does the planning system work? The 2015 Budget in the UK allocated £40m to the 'Internet of Things', which will make places smarter and more dependent on smart technologies (**sustainability**).

The property development industry and investors and providers of finance characterise city challenges differently because they see places as economic units that both generate and consume money. In 2012 the *Financial Times* reported that places with few new start-up businesses needed initiatives to enhance their economic base. Belfast, Sunderland, Stoke-on-Trent, Mansfield, Swansea, Hull, Dundee, Barnsley, Plymouth and Wakefield were all identified as problem places. These are the cities most affected by deindustrialisation. In the cases of Sunderland, Grimsby, Stoke and Swansea, these were places where many businesses were started, but more failed than succeeded.

The property development company Jones Lang LaSalle (JLL) identified three types of city based on the potential performance of the property market (Table 28). The table also lists the buoyant, stable and struggling cities identified by the research group Centre for Cities (it excludes the largest cities). JLL uses the term city to refer to a large urban area. Centre for Cities uses a set of criteria to identify Primary Urban Areas (PUA), which are referred to here as cities.

Table 28 City typologies after Jones Lang LaSalle and the Centre for Cities

Jones Lang LaSalle city types		
Growth leaders	**Potential performers**	**Awaiting lift-off**
Brighton Coventry Derby Portsmouth Reading Solihull Swindon	Bournemouth Cardiff Luton Northampton Oxford Peterborough Southampton Warrington	Bedford Leicester Liverpool Plymouth Sheffield Swansea York
Centre for Cities city types		
Buoyant cities	**Stable cities**	**Struggling cities**
London Milton Keynes Cambridge Reading Crawley Oxford Aldershot Bristol	Bournemouth Portsmouth Northampton Southampton York Leeds Peterborough Preston	Bolton Barnsley Middlesbrough Hull Blackburn Birkenhead Burnley Stoke-on-Trent

All of these typologies affect both the place and the people who live there. Do these places represent themselves in the same way? Is the age structure the same or are the struggling cities older and growing more slowly? Will a middle-aged person in a struggling city view their home place in the same way as a similarly aged person in a stable or buoyant city?

In 2014 the Royal Town Planning Institute (RTPI) noted that most governmental documents that examined the future for cities did not contain maps and therefore lacked an awareness of space and place. As a consequence, policies may not recognise the interconnections and interdependence between places and the impact of decisions on places and the people who live in them.

The New Cities Foundation (www.newcitiesfoundation.org) has a range of case studies that suggest how people and places can respond to the challenges of the twenty-first century.

Exam tip

Always plan any piece of prose during the exam, even for the mini-essays. If you do not finish the essay, the examiner still has the plan to assess.

Challenges where there is overheating

Overheating is a term taken from economics that describes an area where increased demand (in our case, for housing and office space) results in rising prices rather than increased output, so that the supply fails to meet the demand. London is the classic geographical case of a city with the potential to overheat. There are 10 times per annum more new jobs in London than in any other city in the UK. Some drivers of overheating are outlined below:

- **New twenty-first-century employers:** in 2016 the Francis Crick Institute, King's Cross opened in London, becoming the biggest biomedical research centre in the world. It employs 1,500 highly qualified staff due to: (a) expertise in the universities, (b) expertise and patient variety in London hospitals, (c) the appeal of London as a global city for young scientists and (d) the financial and legal expertise needed to support the commercialisation of research. This is just one example of the quaternary industry driving demand for space.
- **Brownfield land:** 10.1 ha of land at White City will soon become the site for another Imperial College research hub and an innovation district to include The Royal College of Art campus. However, finding sites for the ambitions of others is difficult. Shoreditch, Hackney, Stratford and south of the Thames are being rapidly developed for new activities.

Exam tip

Have other examples of named companies that are expanding and contributing to overheating. It often pays to have two examples, because that shows breadth of knowledge.

Drawbacks of overheating

- **Housing:** a potential drawback of growth is housing provision for the 2,000 new people who arrive in London every eight days. Neither these new arrivals nor existing residents in some areas will be highly qualified. 12% of Londoners live in overcrowded conditions. London needs 42,000 new homes a year for a decade, not only for the increasing workforce but also because the size of households rose from 2.35 in 2001 to 2.47 in 2011. House prices have been rising at up to 10% a year, whereas outside of London the rise is 3.1%. Therefore, to get adequate housing, some are being forced to commute larger distances from the South East and beyond. The processes of housing shortages and rapidly rising prices outlined here, together with foreign investors buying up property as an investment, have been labelled 'plutocratisation'. It is estimated that to meet housing demand in the UK over 300,000 new homes a year are needed, increasing the pressure to build on greenfield and greenbelt sites.

- **Transport infrastructure:** this is under strain, even with the building of Crossrail/Elizabeth Line and the completion of London Overground Orbirail (orbital railway). The pressure to expand airport capacity is yet another consequence of the heat of the London economy and the wealth of its society. Projects such as Crossrail/Elizabeth Line are leading to further house price increases close to the stations on the line.
- **Brain drain:** some are of the opinion that overheating in London draws talent away from the rest of the country, with London-based firms dominating the economy of cities beyond the capital. As a consequence, other cities underperform. The 'metropolitan elite' is often criticised in the media, which reflects concerns that London is overheating economically, politically and socially (**difference**, **inequality**, **resilience**, **risk**).
- **Fast-growth cities:** these are cities that perform well on many indicators, and include Cambridge, Milton Keynes, Norwich, Oxford and Swindon. Their economies are strong and productive, with some of the highest GVA per worker, generating over £3,000 per worker per year more than the British average. These areas are attractive as places to live. However, the downsides are transport congestion, shortages of housing, which can be unaffordable, and the problem of those with no qualifications who are unable to find jobs in cities demanding high-level skills.

Challenges of segregation and inequality

A lack of regeneration or an overheating urban area can lead to changes in the socioeconomic characteristics with resultant issues of segregation. A characteristic of cities in the USA in the twentieth century has been growing segregation, a feature that is increasingly evident in British cities. The modern large city is marked by increasingly sharp inequalities, aided by the high earnings of those in the quaternary sector and legal services. The inequality grows with city size. In the 1840s Engels noted segregation in the industrial UK in Manchester, while Disraeli identified the two nations: the rich and the poor. Spatial segregation is a process over time that involves the following:

- **Concentration:** the increasing juxtaposition of similar social and racial groups.
- **Invasion:** the migration of similar groups into an area.
- **Succession:** the replacement of one group by an incoming group, e.g. Somali people replacing Afro-Caribbean people.
- **Flight:** often referred to as 'white flight' due to the departure of the former inhabitants of an area. Between 2001 and 2011, according to the research group Demos, over 620,000 British people left London when the population rose by over a million.

Knowledge check 23

What is the possible link between 'white flight' and changes in the socioeconomic characteristics of some rural areas?

Types of segregation

- **Ethnic:** the concentration of minority ethnic groups, for example the multi-racial population in the Lozells area of Birmingham or Bangladeshi people in Tower Hamlets, London.
- **Class:** the concentration of particular earning and employment groups.
- **Life cycle:** the grouping of people at stages of their lives into specific areas. In London, Hoxton and Clapham have above-average concentrations of single people, whereas Richmond-upon-Thames has an above-average share of older people.

- **Lifestyle:** for example, student areas.
- **Linguistic:** often related to the types above. Immigrant groups with a common language may form a community by living in the same location. As a result they may have limited interaction with English speakers, which may hinder interaction with the wider community.
- **Religious:** concentration of adherents to a religion over time, for example Belfast.

These examples are all taken from Cheshire, J. and Umberti, O. (2014) *London: The Informational Capital* (Particular Books), which contains many maps of various types of segregation.

Factors causing residential segregation

- **Ability to pay for housing — whether to buy or let:** this results from inequalities in income. According to the Joseph Rowntree Foundation income inequality is highest in London but also high in other towns and cities, such as Reading, Bracknell, Guildford, Watford.
- **Availability of housing:** a lack of starter homes for young persons or social housing/housing association properties for low-income households.
- **Gatekeepers:** landlords, estate agents, banks, mortgage companies and 'the bank of Mum and Dad'. By raising rents, often on an annual basis, landlords can re-engineer the social make-up of a neighbourhood. Rents are set at levels determined by demand but also at levels that either exclude or include certain social groups. Mortgages are awarded on the basis of ability to pay both the deposit and the subsequent monthly repayments. More affluent parents can assist with deposits that might enable their children to live in the more desirable areas or have a foot on the ladder towards living in the most suitable neighbourhood for their lifestyle.
- **Demand for housing:** in some places this is caused by gentrification that pushes, for example, Londoners to the outer boroughs and beyond. Overseas investors are also raising the level of demand in London, yet leaving expensive property empty.
- **Threat hypothesis:** segregation is stimulated by perceived and actual threats to the way of life. Besides perceived social threats, the perceived danger of crime often portrayed by the media, such as riots, will deter some people and attract those who need lower housing costs.
- **Marginalisation of workers:** especially in manufacturing due to deindustrialisation and competition with both the more educated and other ethnic groups, who may be immigrants, for new jobs.
- **Government policies towards housing immigrants:** refugees and asylum seekers, for example Kosovan people in Croydon and Somali people in Cardiff. There are also government-defined dispersal areas for asylum seekers that lead to the growth of refugee communities in places such as Bolton, Portsmouth, Rotherham and Swansea.
- **Past government policies:** especially the building of council estates, now called social housing, which pushed lower income households into peripheral estates. Despite the 'right-to-buy' initiative, many of these areas are still marginalised.
- **Immigrant groups:** these often congregate in the area in which they first arrived before spreading to other areas over time. A former Welsh chapel on the Mile End Road, E1, is a representation of a nineteenth-century cluster of migrants from Wales. Bangladeshi people cluster in Tower Hamlets, which has traditionally been

the first home for these migrant communities. Family and friends come to the same neighbourhood, often referred to as chain migration.

- **Affluent households:** these areas attract more affluent people; the greater the wealth of the city expressed as GDP per capita, the more affluent the areas present.
- **Property development industry:** this creates fear by building gated and walled developments.
- **Unforeseen consequences of past 'social' policies:** the following points are made by Professor Danny Dorling of Oxford University. In the second part of the twentieth century a series of policies to address inequalities have given rise to greater inequality and segregation. Educational reforms in the twentieth century have led to a greater body of well-qualified people, both male and female, and to them being segregated by educational outcome as a new elite. In the 1960s full employment meant that all had their basic needs satisfied. However, full employment led to the wages of those in the worst jobs declining while those in the better positions saw salaries rising. The consequence was the beginning of social exclusion as the rich bought more, whereas those in poverty were excluded from being consumers.

  Prejudice against the less fortunate by birth, ability, home location and place in the labour market rose, thus reinforcing segregation. Affluence has led to greed for ever higher salaries to enable people to live a life that equates with their status as, for example, bankers or lawyers wishing to live in the most desirable suburbs of a city. The unfortunate by-product of these trends is that there are those in a state of despair who might live in overcrowded and possibly insanitary conditions in older buildings. Most segregated are the homeless street dwellers and those who depend on food banks.

How can segregation and inequality be reduced in places?

> **'Geographically, with each year that passes, where you live becomes more important.'**

Danny Dorling, *Injustice*, 2010

Income and wealth inequality were noted in the specialised concepts as 'the greatest threat to society today' and a dominant cause of segregation. Segregation emphasises differences between places at the neighbourhood and local scales. People within areas that are different can and do form attachments to the area and become resilient when the status quo is threatened. Therefore, policies to address segregation involve risk and mitigation, and are the responsibility of local and national governments that are politically motivated.

One consequence is that policies and outcomes for localities will vary according to the political ideologies of the decision makers in government and/or private companies at the time of implementation. The building of social housing estates in the last century segregated some people to the fringe and others to high-rise blocks in the inner city. Both of these solutions have resulted in problems that were manifested by, for example, riots in Tottenham and in the Parisian *banlieue* (suburbs) of Grigny. The twenty-first-century response is often to tear down the inner-city towers and replace them with terraces, whose inhabitants have the right to buy. How can we alleviate the problems related to segregation and inequality? The following is a list of potential solutions (note: the list is not in any particular order).

- Policies that tackle the fact that as affluence grows more people are living in poverty.
- Policies that demand that new developments have a mix of housing, including social housing and 'starter homes'.
- **Contact hypothesis**, which states that areas with greater diversity associated with more inter-ethnic contact have lower ethnic animosity, whereas 'white' neighbourhoods with little diversity are unaffected.
- Laws to stop discrimination in the rental market, which was prevalent in the past when signs on doors stated who would not be offered the accommodation to rent (signs in the 1960s drew attention to nationality and skin colour).
- Benefits policies, which were originally designed to reduce inequality.

Summary

- Movements such as Safe Cities and the Transition Towns Network are attempts to alter the social and economic life of urban places.
- Challenges remain despite the efforts of those who rebrand and reimage places.
- Over-successful towns and cities have challenges that result from overheating of economic activity and consequent social change.
- Spatial segregation and inequalities have risen alongside the economic success of urban places.
- Measures to reduce inequalities and spatial segregation are dependent on governmental policies, and they often have unintended effects.

Questions & Answers

About this section

The questions below are typical of the style and structure that you can expect to see in the exam papers. Each question is followed by comments that give some guidance about question interpretation. Student responses are then provided with further examiner comments indicating the strengths and weaknesses of the answer and the number of marks that would be awarded.

When examiners mark your work, they use a grid that gives the maximum available mark for each assessment objective (AO). The mark scheme will have an indication of what should be included in the answer as well as marking guidance for the criteria required to reach the different mark bands.

You should make use of examples where appropriate and reference data to support your answers. You can include sketch maps and diagrams, where relevant. For AS exams the answers are written in the examination booklet with the number of lines indicating the level of detail required. When writing in an answer booklet it is important to number your answers in the same way as the examination paper. If you use an extension page you must make a note, such as 'continued on page…', at the end of the previous page. Remember to number the question on the extension page.

The formats of the different examination papers for this theme are given in the table on page 5.

Question 1 (WJEC AS format)

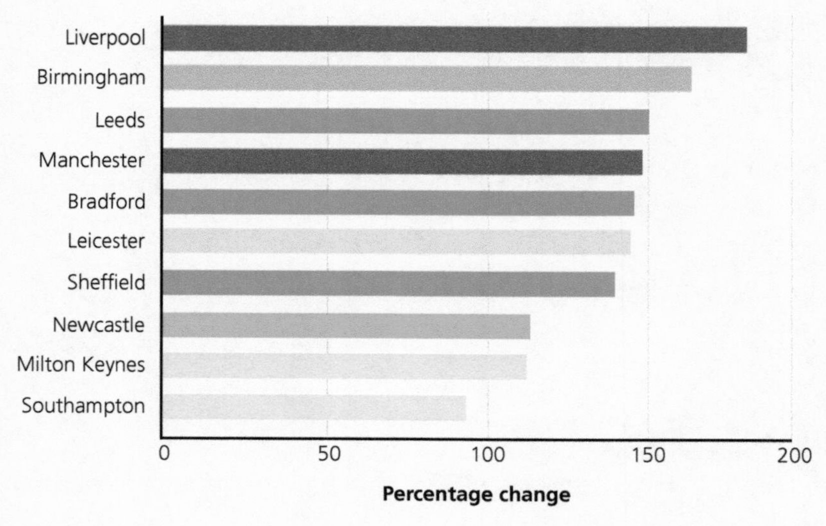

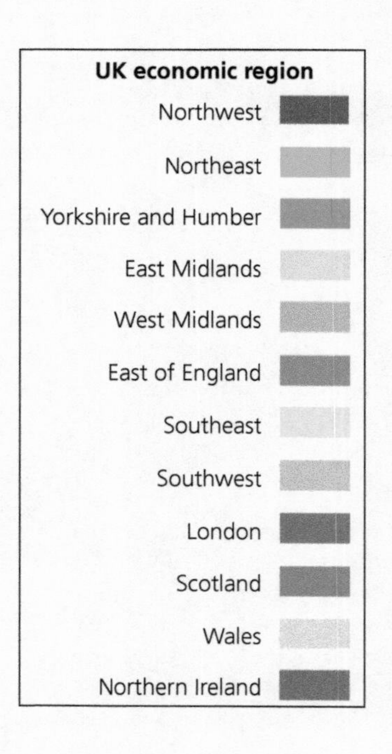

Figure 1 Fastest-growing UK city centre populations 2002–2015

(a) i Use Figure 1 to describe the pattern of fastest-growing UK city centre populations. [3 marks]

The command word 'describe' is targeting AO3, so you have to interpret and analyse the information shown in Figure 1. You must provide factual detail about the characteristics of the data that have been given and identify the distinctive features. All 3 marks will be awarded for your AO3 — your use of skills to interpret and analyse the data.

Student A

Out of the 10 cities, nine have had an increase greater than 100%, with Liverpool being the greatest. There is a concentration in regions of northern England, especially the Northwest and Yorkshire and Humber, with five of the 10 fastest-growing city centre populations. Most are found in old industrial cities where regeneration has occurred. Areas at the edge of the UK such as Northern Ireland, the Southwest and Wales have none of the fastest-growing city centres. Although London has a large city centre, it is not in the top ten.

3/3 marks awarded This answer describes the main patterns concerning the location of the fastest-growing city centres and, to some extent, the rate of growth. There is some use of knowledge when referring to regeneration and the size of London, which is not required in the answer because this information cannot be identified in the data that have been given. Credit is given for identifying areas without the fastest-growing city centres, but it would be better to name them or to use a term such as 'peripheral areas' rather than 'edge of the UK'. 3/3 marks for AO3 have been gained.

ii Justify a suitable alternative method of presenting the data shown in Figure 1. [3 marks]

The command word 'justify' requires you to explain why your choice of data presentation is a suitable method. It is testing AO2 because you are having to apply your knowledge of presentation methods to this subject matter. It is necessary to clearly outline the method you suggest as well as justifying its use. A quick simple sketch of your chosen method could be used for clarification. However, remember that the marks are for justification of its use and not the plotting of data. All 3 marks will be awarded for AO2 as you are applying your knowledge and understanding of data presentation methods.

Student A

A method that could be used is to plot the fastest-growing city centres on a map of the UK. The size of the circle for each city could be drawn to represent the size of the population growth and a scale added to the map. An advantage of this method is that it is very easy to see where in the UK the fastest-growing city centre populations are located.

2/3 marks awarded While this answer suggests a method that is suitable and offers limited justification, it is rather simplified, so only 2/3 marks for AO2 can be awarded. It would be improved by referring to the method as 'located proportional circles' and noting how spatial patterns can be more easily identified if they are located on a map. Areas with no cities in the top 10 can also be easily identified. While highlighting the ease of seeing the location of the cities, the answer fails to mention how the one diagram would have the advantage of being easy to understand, allowing comparisons between cities and regions.

(b) Suggest **one** reason to explain why there are variations in the rate of growth of the city centre populations. [2 marks]

In this question the 2 marks will be awarded for your ability to apply your knowledge and give a plausible reason to explain the trends shown by the graph (AO2). There are a number of possible explanations for the variation in growth rates. The question clearly states that *one* reason should be given, so there is no credit for multiple reasons. There are 2 marks for this question, so the answer requires a little elaboration. However, the command world 'suggest' shows that it does not require a lengthy explanation.

Student A

Many of the cities shown in Figure 1 have central areas that have declined, such as the old traditional industries or areas of docks and warehousing. These have been redeveloped to provide new homes, apartments and jobs, which attract large numbers of young professionals who want to live near the city centre. Cities with no redevelopment may still be in decline, attracting fewer people to live there, or encouraging people to move elsewhere, and so have a slower rate of growth.

2/2 marks awarded This is a concise answer, which suggests and elaborates on one reason only, so gaining the maximum mark of 2/2 for A02.

(c) **Examine the role of the reimaging and regeneration of urban places in creating conflicting perceptions.** [8 marks]

This question requires you to demonstrate your knowledge and understanding of some of the issues that may arise (A01). The command word 'examine' shows that you should consider the interrelationships involved with the issue (A02). Answers should be supported with the use of relevant and up-to-date examples. Where applicable, a learner's 'own place' or examples from fieldwork could be used. Marks are not split evenly between the AOs. 5 marks will be awarded for the demonstration of knowledge and understanding (A01), and 3 marks for how you apply this knowledge to the question (A02).

Student A

When an urban area undergoes reimaging or regeneration this can result in positive or negative perceptions among different parts of the population depending on the impact it has on them. Regeneration can result in run-down areas being improved, with old buildings being demolished or converted into housing. This can improve the appearance of the area and the environment, which may produce a positive perception among locals and visitors. This has happened around the canal basin in central Birmingham.

Reimaged areas can attract modern hi-tech industries, such as in Shoreditch in East London, which in turn attracts young, high-earners to the area. This can lead to a positive perception as an area of growth and a good place to live, which may lead to further regeneration.

The way an area is reimaged or regenerated may attract visitors and tourists. This has happened in Liverpool with the regeneration of the docks area by the River Mersey. The people using the area have a positive perception and perceive it as an attractive place for leisure activities.

However, changes can also result in increasing negative perceptions. People who lived in the area before the regeneration may feel excluded from the jobs and new services that have been created and so do not like the changes. If an area becomes popular with young professionals or tourists this can result in increased house prices and overheating, such as in parts of London. This can create negative perceptions among those who cannot afford to live there. Increasing popularity may also result in a negative perception due to overcrowding and congestion affecting the quality of life.

5/8 marks awarded This answer is a competent attempt to examine the issue of changing perceptions. It highlights how perceptions can be both positive and negative and to some extent relates this to the impact the regeneration can have on different groups of individuals, although it lacks a great deal of detail. This gains 2 of the 3 marks available for A02. While examining what influences perception it would benefit from more factual detail from case studies, which would help it reach the top band of the mark scheme where knowledge of well-developed examples is required. Where examples are used, they are lacking in detail, which impacts on the A01 mark. 3 of the 5 marks available for A01 are awarded.

Question 2 (Eduqas A-level format)

Spearman's rank was used to test the correlation between the two sets of information about the UK's knowledge economy, shown in Figures 2 and 3.

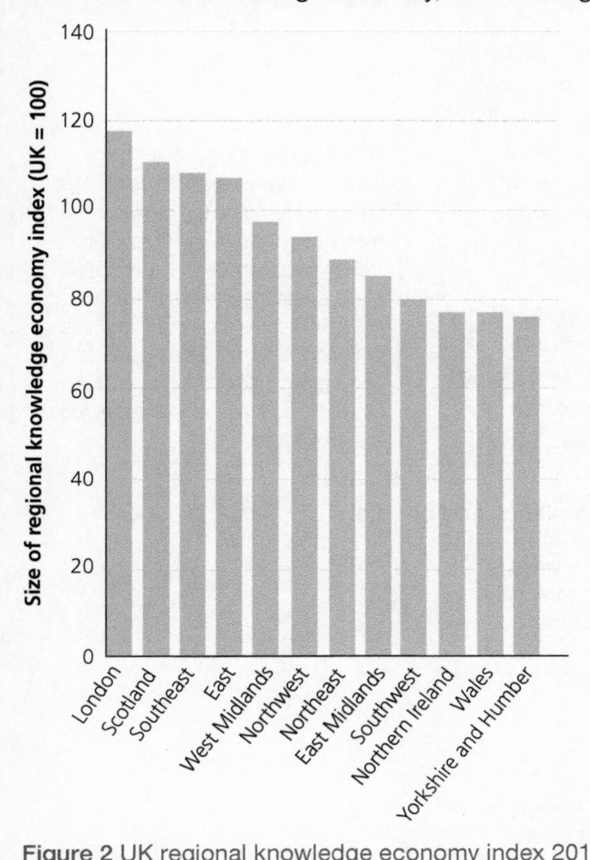

Figure 2 UK regional knowledge economy index 2017

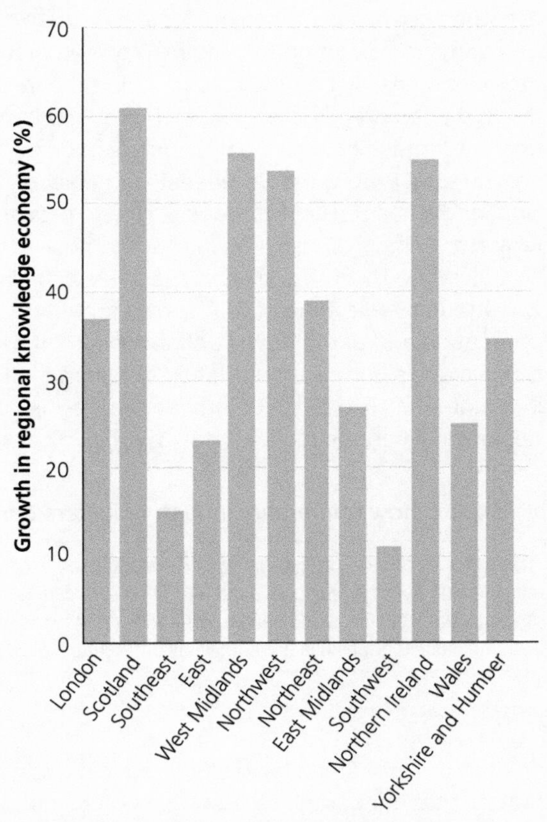

Figure 3 Growth in the regional knowledge economy 2009–2017

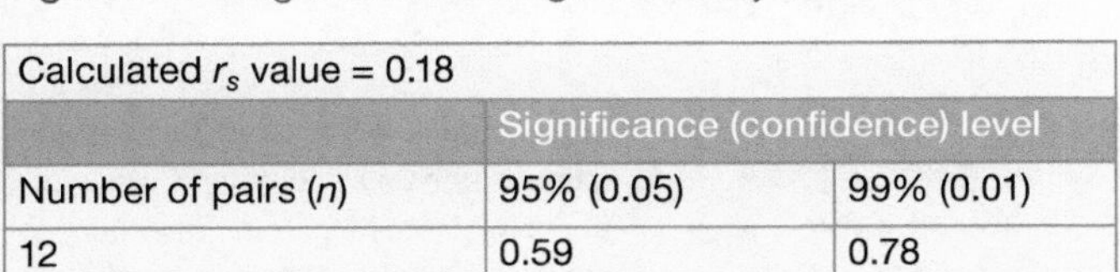

Calculated r_s value = 0.18		
	Significance (confidence) level	
Number of pairs (n)	95% (0.05)	99% (0.01)
12	0.59	0.78

Figure 4 Calculated r_s value and critical values for Spearman's rank correlation coefficient

(a) **Use the information in Figures 2, 3 and 4 to analyse the relationship between the size of the UK's knowledge economy and the growth of the knowledge economy in the economic regions of the UK.** [5 marks]

> You are expected to be familiar with and able to use many geographical skills, including a variety of number and statistical calculations listed in the specification (AO3).
> The command word 'analyse' requires you to identify the essential elements of the relationship. The inclusion of a Spearman's rank test is testing your understanding of the statistical test technique and must be referred to. Evidence from the graphs should be used to support your understanding of the Spearman's rank test. All 5 marks will be awarded for the interpretation and analysis of data (AO3).

Student B

Figure 2 shows that the size of the knowledge economy varies in the different regions of the UK. Only four regions have above the UK average and three of these are located in the Southeast and East of England, including London. Figure 3 shows that the rate of growth has varied greatly between regions, and the fastest growth does not mean the region has a large knowledge economy. Scotland, with the second-largest knowledge economy sector, has the highest rate of growth (61%), while the Southeast with a sector slightly smaller than Scotland's has the second-lowest rate of growth (15%). Although a region may have a high growth rate its knowledge economy may still be below the UK average, as is the case with Northern Ireland. The Spearman's rank value of 0.18 shows that there is only a very weak correlation and at the 95% confidence level it cannot be said that there is any significant correlation between the two sets of data shown in the graphs. The areas of largest knowledge economy have not necessarily had the fastest rates of growth between 2009 and 2017.

5/5 marks awarded This answer analyses the data competently. It clearly highlights the essential elements shown by the graphs and makes links between the two sets of data, quoting examples from the graphs. There is a clear understanding of Spearman's rank and the meaning and significance of the result. This achieves the top band and gains 5/5 of the A03 marks.

(b) Explain how quaternary industry clusters can impact on places and people. [8 marks]

This requires you to demonstrate your knowledge and understanding of the impacts, so is testing A01, for which all 8 marks will be awarded. It is important to cover both aspects, namely places and people, and the varying types of impact, which can be regarded as either positive or negative by different people. Examples should be used.

Student B

Quaternary industries have clustered in certain areas where the factors that encourage clustering, such as nearness to a university and the correct infrastructure, are found. The development of a cluster can improve the global connectivity of the area, which can attract further investment. MNCs may be attracted, producing a global hub. This has happened at Tech City in East London, where Google and Amazon are involved. However, places that do not have suitable locating factors may decline as new businesses, and some established ones, are attracted to areas where clusters are found. Workers attracted to clusters can earn 36% above the UK average. This can result in the gentrification of an area, improving buildings and services, which can change the nature and perception of a place. This has happened in parts of East London, such as Mile End.

The formation of a cluster can increase job opportunities for people in the area, plus the multiplier effect can add an extra five jobs for every one high-tech job. However, for those lacking the required skills, the job opportunities are limited. The demand for property can push up prices and some people will be unable to buy in the area. This can also prevent new start-up businesses from opening and possibly joining the cluster, as has begun to happen at Tech City.

5/8 marks awarded This answer demonstrates a clear understanding of the impact resulting from quaternary industry clusters. It has some well-developed explanations and gives equal balance to positive and negative impacts on both places and people. These features are required for a top-band answer. However, a top answer also requires well-developed examples. In this instance the examples are rather simplistic and when some points about the impact have been made there has been no use of an example to support the statements. 5/8 marks are awarded for the demonstration of knowledge and understanding (A01).

Tertiary services in the cluster area may change to cater for the new workforce and population. This may result in the original or unskilled population becoming socially excluded. People and places outside the cluster will not benefit from any advantages it brings. They may be digitally excluded as there is limited demand to improve and justify the cost of improvements to the digital infrastructure. This has occurred in rural areas such as parts of Northumberland.

Question 3 (Eduqas A-level format)

Assess the impacts of changes in the service economy. [15 marks]

For this question you must demonstrate your knowledge and understanding of the impacts resulting from the changes that may occur in the service economy (A01). The command word 'assess' suggests that there are a number of possible explanations and outcomes. You must apply your knowledge and understanding to evaluate the impacts (A02). It is necessary to give the main explanations and provide justification and evidence to support your points of view.

Where applicable you should use supporting data or evidence, such as case studies. However, do not be tempted to be creative because examiners are familiar with the 'standard' case studies that may appear in textbooks, and can easily check your case study 'facts' on the internet.

A conclusion should be reached but the answer should remain a balanced assessment, which makes it clear that other outcomes are possible depending on the influencing factors that are relevant to a specific location.

In this type of question 10/15 marks will be awarded for A01 and 5/15 marks for A02.

Student A

The service economy, which is made up of retailing, commercial services and leisure and entertainment services, has undergone a number of changes in recent years. These changes include the decline of the service economy in some areas while it has seen growth in others. In some cases, the decline and growth have been brought about by the services moving away from urban centres and changing their location to elsewhere. There are also some places where the types of service that are found have changed significantly. The reasons for the changes are many. However, such changes are likely to have a level of impact on the people and places where they occur. Such impacts may be social and/or economic and can affect different groups of people in different ways.

This introduction clarifies what is considered to be the service economy. It highlights that there have been a number of changes, which can have a variety of different impacts. This sets the scene well for the main part of the essay.

One of the major changes taking place is the decline of services in town and city centres. This includes companies closing stores or ceasing to trade, and services such as banks closing branches. By 2019 over 10% of shops were empty in the UK, and a third of bank branches closed between 2015 and 2019. Many of the areas showing the greatest decline are cities in northern England. The impact of such closures will be felt most by those who are unable to access the alternatives. Traditionally, this is the elderly and children. It is also especially true for those living in small settlements where the closure of a bank, for example, may leave them with no alternative, and also in large cities where the closure of major chain stores can have the greatest effect as this is where they are usually found. Frequently, the elderly may not have access to transport to reach the services they require and may lack the digital skills or confidence to shop or bank online. Consequently, they are excluded from certain services, which can have a negative impact on their quality of life. It has been suggested that such changes to high streets can change shopping for many into a less pleasant functional activity rather than more of a leisure pastime. This highlights how the changes in the service economy can have a negative impact on people by leaving them more socially isolated.

The decline of services in central areas will also have an economic impact. A high street in decline will often lack further investment as companies or individuals are unwilling to expand into the area. A more immediate impact for people living in the area may be unemployment. Between 2016 and 2019 the service economy lost 106,000 jobs in retailing. This will have the greatest impact on women because 70% of all jobs lost were among female employees. This impact may be increased by the multiplier effect as former workers spend less on services, which in turn may be forced to close down.

This gives a clear outline of the decline in services in central urban areas. It highlights how impacts can be both social and economic and considers who might be most impacted by the changes. There is good use of up-to-date figures to highlight the scale of the changes, showing that knowledge is being applied (A02). This demonstration of knowledge will be reflected in the A01 mark.

It would appear that the decline in the service economy in central urban areas has a negative impact socially and economically for many people. Another change has been the renewed demand for the development of out-of-town centres. In 2016 there were planning applications for 24 such centres. The impact of such changes can be an acceleration of the decline of the surrounding town centres as shoppers prefer the ease of shopping and often the free parking that is found at out-of-town centres. Such a change in the service economy over the years will have the same negative impacts as those caused by the decline in the high streets. However, areas close to the new out-of-town centres may feel a positive impact economically from the increase in rates and rents, and the nearby populations may have better access and a better range of shops, as well as possible job opportunities.

The impact of job opportunities can also apply to the fact that many companies are locating in business parks, while leisure activities and stadiums, e.g. Reading FC's Madejski Stadium, are locating on the outskirts of urban areas. While not all are involved in the service economy, many will be, creating a positive impact for those able to take advantage of such moves.

Another reason for the decline in shopping areas has been the growth of internet shopping. This has increased greatly, with 18% of all UK sales being online in 2018. Once again this can have a negative impact on those who rely on physical shops and can result in some being digitally excluded. Frequently, this affects the elderly, who may lack the required computing skills because they were not brought up using them. However, it is estimated that over 80% of the population bought something online during the year. This suggests that for many people the change has had a positive impact because it has increased their access to a larger range of different services.

There has also been an increase in services offering deliveries and click-and-collect services, such as the major supermarkets. For many, such as non-car drivers, those living in areas of poor public transport and those with mobility problems, such services can have a positive impact because they allow these people greater access to the service economy. Economically, the impact of such changes may be negative in as much as the decrease in physical buildings may result in less income from rates and rents. On the positive side, distribution centres, such as Amazon's, do create a large amount of employment, although these are likely to be away from the central areas where many service economy jobs are lost.

This section highlights that the growth in certain areas, combined with changing shopping habits, has mostly negative impacts. However, it also suggests some positive impacts, which demonstrates an understanding that there may be a number of outcomes. Again, data are used to show the extent of the change. This understanding and application of knowledge will be reflected in both the A01 and A02 marks.

In 2017 the amount spent in the UK on retail increased and while most increase was online, there was a 2% increase in stores. Increased disposable income has led to an increased demand for service economy activities. In some areas, including city centres, there has been an increase in shops and services. This is often the case in areas where regeneration has taken place, such as the area around the old docks in Liverpool or the Gas Street Canal basin in Birmingham. However, the new services may reflect the changing demand from the new population in regenerated areas. Many central areas are being populated by a younger, professional and affluent population. Services such as restaurants, coffee shops and barbers often increase in such areas, along with leisure services like more exclusive gyms.

While having a positive impact for the new population as the services meet their needs, for many of the older, original population the impacts are once again frequently negative because their needs are not necessarily met. Much of the younger population will have adapted to an increasingly cashless, online society, so there is little need for banks, post offices or access to cash via ATM machines. Some shops and services will now no longer accept cash.

For those who still rely on using cash, the lack of access to it can make life more difficult. Also, the cost and types of newer services may result in some members of the population becoming socially isolated.

In some places, regeneration or rebranding has resulted in an increase in tourism and the services such as shops have changed to meet the demands of the tourists. For example, in Port Isaac in Cornwall, used for the filming of the television show *Doc Martin*, the bank and some other shops have changed to selling souvenirs of the show. While the tourist influx has a positive impact in bringing in money, the resident population has suffered from a decrease in available services. This is an example of how a small village has been impacted and how the changes in the service economy do not just affect urban centres. In rural areas the closure of shops, banks and pubs may have a much greater social impact because there are fewer or no alternatives nearby, thus increasing social isolation and exclusion. This is especially true because much of the rural population is ageing and tends to suffer more from the negative impacts.

The changes occurring in the service economy can be seen to have had both positive and negative impacts on people. However, the type and extent of the impact are dependent upon may factors including age, mobility, wealth, digital awareness and accessibility, willingness to adapt, and geographic location. In the future, as the elderly members of the population become those who have been used to internet shopping and mobile banking, it may be that such changes in the service economy have much less of a negative impact on many people. The ability to be able to access services online may also reduce the impact in rural areas, especially those that at the moment do not benefit from fast broadband.

This essay has now extended its assessment by demonstrating that changes in the types of service can have an impact in different places. There is some mention of examples to support the findings (AO1).

This conclusion summarises the findings of the essay, highlighting that many factors can affect the type of impact. It also demonstrates understanding (AO1) by suggesting that future events might change the extent of present-day impacts.

13/15 marks awarded This answer demonstrates good knowledge and understanding of the impacts created by the changes that have occurred in the service economy. It outlines the variety of types and locations of changes and at each stage of the essay links the type of change to the impact it may have. Ideas are supported by factual details, although more use could be made of actual examples (possibly using data collected on fieldwork) to ensure high marks for AO1. However, there is enough evidence to reach the top band for AO1, gaining 9/10 of the available marks. Application of knowledge has produced a well-balanced, detailed assessment, needed for the top band of AO2, gaining 4/5 marks. There is a clear structure with an introduction and conclusion, and the main body of the essay clearly assesses the types and extent of any impacts that may occur.

Student B answer

One of the changes that has taken place in the service economy recently has been the closure of many shops on the high streets throughout the UK. The service economy was traditionally located in city centres as this allowed the shops and entertainments to be accessible to a large population. Many tertiary industries also developed here to provide services for the industries that were found there.

This is a brief introduction which concentrates too much on why service industries are found in city centres. It fails to say what is being meant by the term service economy and does little to introduce the idea of assessing the type and extent of the impacts that may occur. It therefore does little to help gain marks for AO1 or AO2.

Competition from the internet, with more people shopping online, has resulted in many shops being unable to make a profit. This has led to a number of chain stores, such as Debenhams and New Look, closing many stores while others like Toys R Us and House of Fraser have closed completely. Often these shops were 'anchors', which attracted people into the shopping centre. As a result of their closure the footfall in the area has declined resulting in the remaining shops doing less business and consequently closing. At the same time traffic congestion, congestion charging and the difficulty and cost of parking, have also dissuaded people from using town centres, thus adding to the problem and resulting in more shop closures. In 2019 there were over 1,200 more shop closures than openings in the UK. There are also changes in the types of store that are opening. In some areas new shops are coffee shops and restaurants, while in others shops are being replaced by charity shops or those selling cheap goods. Frequently the environment can appear run down, which discourages people from going shopping there and also other businesses from opening.

This very descriptive section outlines why changes are taking place. There are hints at what impacts these changes may have, but these are superficial, and it is the examiner who is having to make the link rather than being told it. There is some evidence of knowledge for A01 but little evidence of application of the knowledge which will limit the A02 mark.

Another change that is taking place is tertiary activities moving to out-of-town locations. Offices are locating on business parks where there is less traffic congestion and a more pleasant environment. The largest in the UK is the Cobalt Business Park in Tyneside where 14,000 people work. Similarly, out-of-town shopping centres are being developed again. Shops are attracted by the space and the ease of receiving goods, while shoppers are attracted by the variety of stores in one place and usually plentiful free parking. A number of stadiums have also moved location, such as Reading FC, which moved to a site built on an old landfill site outside of the city.

A number of cities in the UK are being affected by reurbanisation, with people moving into the city centres, often in converted industrial buildings. Often the people moving into such areas are affluent, young professionals and the area reflects this change by gentrification, where poorer-quality areas are improved. Large parts of East London, such as Wapping and Shoreditch, have been transformed. With a good disposable income there is a demand for services, and many have moved into such areas to cater for the demand. These services are places such as coffee shops, restaurants and exclusive leisure centres or gyms as opposed to shops selling goods.

Again, this section describes the changes in the service economy, giving some explanation as to why they have occurred. It does not mention the impacts the changes have had, so does not help in the assessment of them. This lack of application of knowledge will limit the A02 mark.

It can be seen, therefore, that the service economy is changing, with a decline in some places and growth in others, albeit with a change of services being offered. All these changes can have an impact on the area concerned. Some people will find it harder to access services if they cannot travel to out-of-town locations or if the new services moving into the area do not meet their requirements. Changes can also impact on the environment, with land being used for new centres

It is important to realise that in essays with the command word 'assess' (and also 'discuss' or 'evaluate'), the process should continue throughout the essay and not be left to a final conclusion. Each paragraph or section should make a valid point that is then related back to the title to show how it can contribute to the final conclusions that are drawn. This highlights to the examiner that you are applying your knowledge (A02) to produce a focused assessment rather than writing all you know about the topic.

and increased traffic flows to get to them, while fewer visitors to city-centre services may result in improved air quality. Some people may also lose their jobs due to shop closures, while there may be more jobs available when the out-of-town centres open.

All in all, the changes in the service economy will have impacts on many people because they will affect their lives. While for some they will make life harder, others may find the changes have little effect on how they live their lives.

This conclusion gives a limited assessment of the impacts. Some of the impacts mentioned concern the environment. However, the specification (Section 1.3.5) limits the study of the service economy to its social and economic impacts, clearly referring to impacts on learners' own lives and the lives of others. Therefore, any mention of the impacts on the environment must be related to how this affects people.

5/15 marks awarded This answer demonstrates some knowledge of the changes to the service economy. However, it is mostly descriptive about these changes and there is little assessment of the impacts, including the types of impact and who is affected. There is little use of factual evidence to support any findings. It is therefore making little attempt to answer the question that has been set. The structure of the essay is simplistic, with a limited introduction and most of the relevant assessment of the impacts being left to the final concluding sections. The demonstration of limited knowledge that is relevant to the question means it is a band 1 answer for A01, gaining 3/10 marks. Some of the knowledge is applied to give a partial assessment, allowing 2/5 marks to be awarded for A02.

Question 4 (Eduqas A-level format)

Evaluate the role of counterurbanisation in creating challenges for rural communities. [15 marks]

With this question you need to demonstrate the ability to develop a sustained line of reasoning. Your answer should be coherent, relevant and logically structured. It should be substantiated with suitable examples. Two-thirds of the marks are awarded on your ability to demonstrate detailed and accurate knowledge and understanding relevant to the question (A01). The remaining marks are awarded based on how you apply your knowledge to construct a well-developed evaluation that is supported by evidence (A02). As with the previous question, remember the importance of accurate, relevant and up-to-date evidence and case studies.

Student A

Counterurbanisation is the movement of people and sometimes businesses from urban to rural areas. As a result, the size of the rural population in the UK has increased since the 1980s. Many rural areas are facing social and economic challenges, which could possibly be as a result of counterurbanisation.

This is a suitable, succinct introduction. It defines the key term and highlights that the challenges can be both social and economic in nature.

One of the challenges for rural communities is the issue of higher house prices. Increasing demand for homes as a result of counterurbanisation has resulted in UK rural house prices doubling in the last 10 years and, on average, increasing at a faster rate than urban areas. Many people moving into rural areas may have higher-earning jobs or have sold a house in an area where prices are high, such as London or Southeast England. This means they can afford to pay higher prices in rural areas. However, this creates a challenge for people in rural communities, who often have lower-paid jobs, and the young trying to get on the property ladder, who are priced out of the local market and so may be forced to move out of the area. [a]

People moving into rural areas may still commute to nearby urban areas for work and frequently make use of the services in the urban area. Rural services may close due to the lack of use. For example, up to 500 village shops close each year. [b] Over a third of those moving to local areas do so for retirement. In Wales 25% of the population in rural areas are over 65. [c] Such populations do not have young children so there are not enough pupils to keep village schools open. Closure of such local services may therefore cause challenges for members of the rural population who do not have access to services because they are unable to drive or have no public transport. At the same time, an increase in the elderly population may pose the challenge of providing services such as medical and health care.

This section shows how counterurbanisation can contribute to the challenges for rural communities. The use of an example to support the point being made at [a] would have been beneficial in demonstrating knowledge for A01. The use of figures at [b] and [c] demonstrates knowledge of the subject, helping to gain marks for A01.

However, some challenges may be eased by counterurbanisation. Increasingly, services such as libraries and public transport are run by volunteers and newly retired people who move to rural areas and will often take on these roles. It is also possible to limit counterurbanisation by restricting the sale of homes to locals only. For example, community land trusts, such as the one in St Minver in Cornwall, can manage homes, keeping them affordable for locals.

Showing how counterurbanisation can ease challenges and how its impacts can be reduced helps to demonstrate a good level of understanding and to produce a balanced evaluation. This application of the knowledge will assist in gaining A02 marks.

Some rural communities may be little affected by counterurbanisation, but may still face a number of challenges. The effects of technical developments, foreign competition, foreign labour and EU quotas have helped decrease primary employment to 1%. Many of the jobs that remain tend to be lower paid. As a result, many of the younger population leave rural areas in search of work and higher-paid jobs, leaving an elderly population and the resultant challenges this causes.

As a result of the internet, the growth of online shopping and the development of supermarkets increasing their delivery areas, many rural services such as shops, banks and libraries are not economically viable and close, or shops change to cater for tourists. Access to services therefore becomes a challenge for the rural community without any contribution from counterurbanisation. Access to services may also be a challenge for rural communities due to the level of broadband provision. While it is improving, there are areas, such as the very remote parts of Wales and Scotland, where internet access is limited, which in turn results in the population being digitally excluded. It can also prevent investment in the area by businesses. Again, these challenges are not the result of counterurbanisation.

This section demonstrates knowledge and understanding of further challenges for rural communities, highlighting how counterurbanisation has had little or no role to play. This has helped produce a balanced evaluation of the role, demonstrating application of knowledge (A02). The use of a case study at [d] provides evidence to support the point that is being made (A01).

Regeneration and reimaging can have an impact on rural areas by changing the perception of the place and increasing tourist numbers. Port Isaac in Cornwall is used for filming the television show *Doc Martin*. This has created a number of challenges for the local population, such as dealing with large numbers of visitors and traffic congestion. House prices have quadrupled, pricing out locals, and the area has become popular for people buying second homes. Services have changed to cater for the visitors, with the opening of souvenir shops. **d** Rural communities such as this therefore have a number of challenges that are not related to counterurbanisation.

Rural communities can face many challenges and it can be seen that counterurbanisation has contributed to some of them. However, in areas where counterurbanisation has been limited or non-existent, the rural community may still face many challenges, so counterurbanisation cannot be totally to blame. In fact, it may also be the case that counterurbanisation helps solve the challenge of maintaining services in rural areas.

This concise conclusion provides a brief summary of the evaluation findings without repeating much that has previously been written.

11/15 marks awarded This answer demonstrates knowledge and understanding of the challenges faced by rural communities and the role counterurbanisation can play in these challenges. There is some factual detail to help support the ideas and some use is made of examples. These tend to be a little superficial in detail, which impacts on the AO1 mark — it just reaches band 1. The application of knowledge has produced a well-developed evaluation of the role of counterurbanisation, which gains AO2 marks. The answer clearly explains how challenges exist without counterurbanisation, with some evidence to support this. There is a good structure, with an introduction and conclusion. As a result, 7/10 marks have been awarded for AO1 and 4/5 marks for AO2.

Student B

Many people may decide to move from urban to rural areas in order to improve their quality of life. Rural areas surrounding large urban areas are often in the greatest demand as well as other areas that are perceived to be attractive places to live. Others may decide to buy a second home in an area. **a** Both of these may create challenges for people who live in rural areas.

This introductory paragraph does not set the scene for the rest of the answer and is not always relevant to the question, for example at **a**, where student B has failed to separate the issues of second home ownership and counterurbanisation. This shows a lack of understanding of the subject and limited application, which will affect the AO1 and AO2 marks.

The people moving to the countryside often have higher-earning jobs and can afford to pay higher prices for homes. As more want to move to the countryside the increased demand for homes can push up the house prices. Locals who are unable to afford the prices may not be able to live in the area and so are forced to move away.

Many people who move make limited use of the local services as they often work and use the services in the nearest town. Second homes often result in villages being deserted during the week and so local services are not used. The newcomers are usually elderly, so do not have children. The result is there is not enough demand for the local services such as shops and schools, so they close down, leaving the rural community with little or no services. This can be a problem for those who are unable to drive, especially if there is no public transport.

Another challenge facing rural communities is employment. The use of machinery such as tractors has reduced the number of jobs in farming. b Most of the remaining jobs are low paid. With house prices increased by counterurbanisation, many are unable to afford a home, so the young move to urban areas to look for higher-paid work.

Sometimes counterurbanisation can cause conflicts in rural areas. Newcomers may want to make changes that the locals do not agree with. For example, they may complain about the noise from the church bells. There may be a demand to build new homes that the locals do not want because they feel they will change and spoil the area.

As can be seen, in rural areas that have been affected by counterurbanisation it can cause or contribute to a number of challenges for the rural community.

This makes some valid points, but these need examples or factual data to provide evidence to support them, which would help attain a higher band for A01 on the mark scheme.

There are some rather simplistic statements here, such as at b, which demonstrate a basic level of knowledge and do little to evaluate the role of counterurbanisation and so shows limited application of knowledge (A02)..

This superficial conclusion adds little to the evaluation of the role of counterurbanisation.

6/15 marks awarded This answer demonstrates knowledge of the role of counterurbanisation but it is simplistic and rather descriptive of the effects. It also refers to the issue of second homes, which is not the same as counterurbanisation. It is lacking factual evidence, and no use is made of examples to support the statements. There is just enough evidence to reach band 2 on A01. While it mentions how counterurbanisation can influence challenges, it does not evaluate the importance of its role. The answer does not provide a well-developed evaluation because it does not discuss how challenges may still be present without the impact of counterurbanisation. This limited application of knowledge allows a partial evaluation and just reaches band 2 for A02. This answer gains 4/10 marks for A01 and 2/5 marks for A02.

Knowledge check answers

1 Increases in wealth and leisure time have resulted in a large proportion of the Lake District population earning their income in the service sector. However, many of the jobs are low paid and possibly seasonal. Increasing visitor numbers have resulted in changes to services as tourist-oriented services increase in the area. The difficulties of farming have seen the primary sector decrease, with farms diversifying into tourism. More affluent people are able to afford second homes. This prices out locals and the young, who leave the area, leaving behind an ageing population. As the service sector in the form of tourism dominates, the influx of visitors can result in increasing issues, such as traffic congestion.

2 There are many examples, including:

- Gateshead — Angel of the North
- Newcastle — Tyne bridges
- Liverpool — Liver Building
- Blackpool — Blackpool Tower
- Edinburgh — Castle
- London — many, including the London Eye, Big Ben, St Paul's Cathedral, Tower Bridge

3 Globalisation has resulted in many shops, especially food outlets, being controlled by MNCs based abroad. Identical outlets in different places means town centres are losing their individuality.

4 See table below.

5 Formal — Office for National Statistics data, tourist organisation promotional material

Informal factual — local historian's account of a past event

Informal non-factual — television drama, literature

Informal opinion — social media, graffiti

6 Perception may be influenced by a person's age, gender, socioeconomic status and socio-cultural position, as well as information that they may receive from formal and informal sources. It will also be influenced by their direct and indirect experience of the location, possibly in the past.

7 An unrealistic aim for a place and its people to be in a state of perfection. Philanthropic industrialists in the nineteenth century, such as Rowntree, Salt and Owen, all attempted to create utopian settlements for their workers. Howard's Garden City movement attempted to harness both the benefits of urban and rural living in Letchworth. Harlow New Town was a socialist attempt to plan a utopian settlement.

8 GVA is the value of the goods and services produced in an area or by an industry. It shows the value produced after deducting the costs of production. GVA per person measures the contribution to the economy of each individual in the region.

9 Some cities may have a range of activities, offering very different wages. York may have highly paid jobs associated with the university and also a large workforce in the low-paid service sector catering for the tourism industry. Coastal places may have an elderly population and therefore a large care sector, but also higher-paid tertiary employment.

10 A test on the data for % no qualifications and the IMD rank give a result of 0.63. This is statistically significant, showing there is a positive relationship — areas with a higher level of deprivation are more likely to have a higher level of % no qualifications. The result of 0.81 shows there is a strong relationship between overcrowding and deprivation, with deprivation to be highly likely in overcrowded areas. The result of –0.18 is not statistically significant, showing that there is hardly any relationship between the amount of terraced housing and the level of deprivation.

Table for Knowledge check 4

Group/ organisation	Engagement with harbour	How harbour locality is perceived	Experiences that influence perception of place
Local residents	Chose to live or have always lived in settlements	Under pressure	Extensions to existing villages, refurbished houses and second homes that increase the urban character of the settlements
Second-home owners	Chose to invest in property	Desirable place to visit with good amenities	Able to buy or even build a property; use of wealth to outbid local population; perceive it as a rural place, especially if resident in London or other major city
Yacht clubs	Year round; moorings, boat parks	Too many boats might spoil the water environment	Getting to water and onto it; quality of sailing environment
Parish Council	NIMBYism; want it to remain as it was in the past; large houses	Beauty, tranquillity, place of leisure; pressure on parking, waste disposal	Litter on shores; developers and redevelopment of existing properties
Conservancy	The environmental value of the waters — promotional; coastal protection	Home for wildlife; wintering grounds for birds	Bird watching
Farmers	Land has footpaths; fear of erosion from rising sea levels	Area into which land drains	Behaviour of walkers, litter

11 Advantages: just shows the city and indicates numbers of jobs. A good initial impression of where employment is located.

Disadvantages: areas other than centre and EZ are not named. There is no scale for the blocks, and there is no distance scale.

12 Many of the traditional 'anchor' businesses such as banks and post offices have closed due to the increasing use of online services. Other shops closing are those that offer a product, for example clothing. The closed shops are mostly being replaced by businesses that offer a service or experience rather than a product. These are things that cannot be obtained online. Convenience stores have increased, reflecting a change of marketing direction by some of the major supermarkets as a result of changing lifestyles and shopping habits.

13 The diagram allows some comparison between the different occupation groups both within each city and also between the two cities. However, there is nothing to say if the figures for each occupation group are percentages. Bars of different length but of similar value (e.g. managers, directors and senior officials) may reflect a different population size but do not highlight the differences and similarities between the two cities. If the main aim is to highlight the differences in the proportion of the populations employed in each sector then bars drawn using the same scale would be beneficial. Aligning the bars on the left could ease comparison. Alternatively, pie charts using the percentage figures, with the sizes of the circles representing the different population sizes of the two cities could be used.

14 Peterborough has gained numbers employed in caring, elementary occupations and a few in management and the professions. The positive changes are similar in Stoke-on-Trent, although the rise in high-waged positions is greater. In both cities the intermediate-waged employees have declined by up to 20%. The trend is one in which both high-waged and low-waged sectors are growing, but it is the low-waged service sector that is increasing fastest. Changes from a manufacturing base to a service industry base are reflected in the graph.

15 Many more of the top 10 cities are located in southern England and the Midlands, whereas only two of the lowest are found there. Most of the lowest are found in northern England where none of the top 10 is located. Scotland has two in the top 10 and one in the lowest group. None is found in Wales.

16 The knowledge economy relies upon research, innovation and technical developments. Areas with high levels of involvement in the knowledge economy activities will therefore be involved in many new developments, which they would wish to protect from the competition by means of a patent. Publicly funded research can also be patented, adding to the number.

17 The cities of the South are strongly represented, as are the capitals, but not Cardiff, which is subsumed into South Wales. There are some deindustrialised places, such as Liverpool, Greater Manchester, Hull and Sheffield, where companies have been able to start up. Research centres such as Oxford, Cambridge and Edinburgh have attracted digital companies.

The data are a mix of cities and regions and so are not always comparable. The data could be grouped into cities and regions. Also, the data do not give a baseline figure for any place. Many cities would have had a large number of companies in 2010.

18 The data are for the period 2008–12. They distinguish rural, less sparse and sparse (remote) rural. The fastest rise in jobs is in health/social work/education/public, but not in the more remote areas. Mining is in decline and agriculture shows a small percentage increase. Overall, it is the rural and sparse regions where there is growth and the less sparse areas where the number of jobs is in decline.

19 Popularity and interest might be short lived and decline rapidly once the series has finished. Speculative rebranding based on a new show may fail if the show is not popular. Alternatively, a sudden increase in tourism may cause issues and influence quality of life for locals.

20 This question expects you to interpret opinions from the facts in the summary. A project cannot keep going if it is dependent on continuous external funding. Many of the attractions cater for a specifically narrow clientele, for example young, adventure-seeking people. Some may suggest that there are more pressing issues to be addressed. The last bullet point hints at social and health problems. Allotments might be viewed favourably by the older generation, as will smart energy, but access to services and transport to towns are not mentioned. Do the schemes retain the population or are the young leaving for university and higher-paid employment?

21 The population living in rural areas is ageing — more are 60–74 and over 75 in 2017 compared with 2001. The number of children aged 0–14 has declined, as has the number in the child-rearing age group, 30–44. These changes have implications for care and school places. Figure 33 only refers to the proportion in the working-age groups and early retirement. It is also about 'usual residents' and not second homes. Almost one-third are in the highest wage-earning groups. Skilled trades mainly involve agricultural, forestry and quarrying workers. Middle earners are almost half of the earners. The data support the view that rural areas are very much regions of well-paid employment and relative affluence, although about 1 in 5 are among the low-paid categories. The impact of counterurbanisation is evident. The resulting issues relate to the impact on house prices, the use of services elsewhere, care and access to care, and second homes.

22 Rebranding may result in increased visitor numbers, causing issues such as increased traffic congestion and demand for parking, and visitors not respecting locals' privacy. Local services such as shops may change to cater more for visitors, thus limiting services for residents. The rebranding may attract a younger and/or more wealthy population, which could socially exclude the original local people. In turn, services could change to meet the new population's needs. New businesses may require skills that the original, older residents do not possess. Rising house prices may encourage locals to leave.

23 The people who take part in 'white flight' tend to be the older, more affluent members of the original population who can afford to leave and may no longer have commitments in the area. Retired people increasingly move to more rural locations, adding to the ageing population. Higher-paid individuals can afford to move to more popular rural locations and commute to work in the urban area. This adds to the more affluent nature of many rural populations.

I

J

K

L

M

N

O